WHERE THE F**K IS BLÖNDUÓS?
DRIVING AND SURVIVING A WINTER IN ICELAND

Emma Strandberg

Published by New Generation Publishing in 2023

Copyright © Emma Strandberg 2023

First Edition

The author asserts the moral right under the Copyright, Designs and Patents Act 1988 to be identified as the author of this work.

All Rights reserved. No part of this publication may be reproduced, stored in a retrieval system or transmitted, in any form or by any means without the prior consent of the author, nor be otherwise circulated in any form of binding or cover other than that which it is published and without a similar condition being imposed on the subsequent purchaser.

Paperback ISBN: 978-1-80369-743-7
Ebook ISBN: 978-1-80369-744-4

www.newgeneration-publishing.com
New Generation Publishing

"There are few things more powerful than destiny, few things fiercer than fate. I am therefore going to Iceland, more because of destiny than out of any personal desire."
Taken from the Vatnsdœla Saga

Contents

LONGHOPE	1
SURTSEY	3
BLÖNDUÓS	10
PLANNING	14
THE JOURNEY	26
THE INTERIOR!	40
GRAPEVINE	54
LEARNING TO KNIT	62
AROMATHERAPY	66
SKAGASTRÖND	70
MANUS I MANO	76
RÉTTIR	80
SOUTH COAST	85
HOME ISLAND	107
THE LIGHTS	120
ICESAR	129
HOT POTS	135
CABIN FEVER	143
TWO'S COMPANY	169
THE SITUATION	180
I COULD EAT A HORSE	185
BLESS BLESS BLÖNDUÓS	191
THE FINAL LEG	199
PS: SURVIVAL	205
PPS: ESSENTIAL ITEMS	207
PPPS: RULES OF THE ROAD	210

LONGHOPE

Experiences throughout life form us. My earliest memory was of a lifeboat disaster. The fourteen-metre Watson-class wooden lifeboat was no match for a force 9 gale, coupled with a spring tide. When the lifeboat and crew left port, conditions were atrocious, with near zero visibility and eighteen-metre-high waves. The officials investigating the disaster predicted the freak wave that overturned her could have been as much as thirty metres. It wasn't the first time the lifeboat from our island had been called out on such a night, but this time it was different. No one returned home.

Our small community was rocked by the tragedy. More than 50 years later, the memories lived on. Those lives lost would never be forgotten, their courage never cease to be respected.

Our family attended the wake and placed a plastic flower arrangement out of respect, gathering it up several days later, washing it and putting it away where it lived in a pillowcase under my bed.

As a child, my vivid imagination nurtured nightmares A wall of water, the weight and pressure pushing me downwards against my will as my mouth opened in a silent scream. My lungs burned. My body spun through a bottomless vortex to eternal nothingness. My world as I knew it turned black. I woke countless nights soaking wet and gasping for breath from that same recurring dream. The fear of water never left me.

My childhood was turbulent and not a particularly happy one. Despite our entire community losing a part of itself that Monday night, 17 March 1969, my mother's life continued unchanged. Events had already made her who she was.

SURTSEY

1963 dominated news headlines in various ways. President John F Kennedy was assassinated. The Polio Campaign saw widespread vaccinations across the UK. The Soviet Union shuttle – "Vostok 6" – carried the first woman into space, and an unexpected event took place in the north Atlantic Sea. The actual event might have spanned a relatively short time, but in our unhappy home, our mother's obsession with it continued for years. That event was Surtsey.

Named after a fire demon in Norse mythology, Surtsey was an unexpected eruption – just as my own birth was unplanned and something of an interruption. Iceland adjusted to having another island just as my family adjusted to having me.

The day was Thursday. As the explosion took place on the seabed 20 miles southwest of the Icelandic Westman Islands, it developed, transformed into volcanic ash, subsequently cooled, and within 24 hours, land was visible above the water where before there had been nothing. A new island was born. It continued to erupt and expand, growing for almost four years. Surtsey was only the second of three volcanic islands of any size to have erupted from the sea and survive for more than a few months. Its distant cousins were Anak Krakatua (child of Krakatoa) in Indonesia and Hunga Tonga and Hunga Ha'apai, the two visible outer rims of the caldera in the southwest Pacific Ocean off the coast of Tonga. As recently as January 2022, Hunga Tonga again suffered a huge submarine eruption separating the "islands" and significantly reducing them in size while blanketing the neighbouring inhabited island of Tonga in thick ash and cutting off all Internet and telephone contact to the outside world for days while contaminating drinking water through the subsequent tsunami waves and ash fallout. The blast was heard some 1400 miles away in New Zealand.

Before her marriage, my mother's own career had been interesting and engaging, allowing her financial independence and the opportunity to travel. Upon marrying my father, she had every reason to expect she would continue to work and enjoy her hard-earned status. A short romance filled with promises in turn led to violence and bitter regrets. Having married too soon after meeting, they set off on a honeymoon that staggeringly stretched 18 months via Madeira, Beirut, the Middle East, and Canada. Upon returning home, and against her wishes, they took over a large, cold property on a remote island off the coast of Scotland. Father, an old school breed, insisted it was the ideal place to bring up children, at least until they were of an age for boarding school. When that time came, Grace could, if she so wished, resume her career. Her family were wary, but given that this was the 1960s, they respected her choices and allowed her to make her own mistakes.

Quite out of the blue one morning, Grace received an international telephone call that had come through to the small post office on our island. An airline ticket would be arriving by post from Icelandair, departing Glasgow for Reykjavik. Her husband, Dow, would be there to meet her. He had been seconded to Iceland some weeks before and wanted Grace to come and visit. Following the call, she was elated to be travelling anywhere that took her away from the daily routine of life on her small island. Despite being heavily pregnant with my eldest sister, Grace didn't hesitate for a moment. She packed her bags, ready to leave in just over a week's time. She prayed her flight would not be cancelled.

It was a twist of fate when two days before setting off on her journey, the news broke of an eruption off the south coast. Her father sent her a telegram begging her not to travel. She ignored him, instead travelling to Glasgow and disguising her pregnancy with loose clothing as she boarded the plane.

On the Icelandair DC-6 flight to Keflavik International Airport, there was talk of little else. In contrast to the bleak Scottish island, starved of company and excitement, our mother found herself in the middle of a historically newsworthy event, and she was captivated. As the flight neared the coast of Iceland, she saw the dots of the Westman Islands beneath her through the grubby window, and in the distance, the plume of smoke that was Surtsey.

On arrival, her optimism at meeting Dow was short-lived. No longer was he the charming, romantic, and attentive gentleman she had married. Instead, she was met by a dipsomaniac whose manners had faded and who had gotten into significant debt. When his demands for money were met with objections, he thought little of using his fists to persuade Grace to part with her bankbook and even her jewellery – which he was instantly able to sell. She, not utterly victimised, stood up for herself physically as best she could. Following a short trip to the local hospital (Landspitalisjukhus) to stitch a badly cut head, she checked into a small guest house and decided to make the most of her situation before the 21-day open plane ticket home expired. She'd be damned if she was leaving a day before she had to.

Having dismissed her violent start, she chose to spend her days in Iceland wisely, talking to anyone she could, digesting every small detail they revealed to her and reading from cover to cover every Icelandic newspaper she could find, including Morgunblaðið and The Nation's Will (Þjóðviljinn). She took photographs by the score for later development. She filled notebook after notebook with her distinctive, beautiful handwriting. During her career, our mother had made several newspaper and media contacts, including befriending a talented journalist named Renton Laidlaw, with whom she corresponded. She wasted no time in exchanging copy and film and even earned a little money for which she was grateful. She made no attempt to contact her husband and temporarily dismissed thoughts of home. Instead, she concentrated on the fact that she was in Iceland, and history was in the making. Surtsey erupting was a gift from God, and Grace was captivated. For those precious few weeks, she was the happiest she had been in a long time. On her last morning in Reykjavik, Grace had been walking past a bookshop when a shiny object caught her eye in the pavement's fresh slush. She picked it up. It was a medallion of Saint Catherine of Alexandria. She called to a gentleman walking a short distance ahead and enquired whether he had dropped it. He had not. With no one else nearby, they discussed what to do and, deciding it probably wasn't valuable, both agreed that she should keep it. She explained to him that her father had been based in Alexandria during the war and wrote home fortnightly from the Cecil Hotel where the Officers dined. For Grace, this find was all the proof she needed that her father had been right about many things. Sadly, she never got to tell him, as he passed away from a skull fracture shortly before Christmas that same year.

Conversation came easily, and the two strangers exchanged stories for some hours and became friends. They walked, talked, laughed and by the time they said their goodbyes, both agreed this small amulet and indeed Surtsey had cemented their friendship forever. His name was Jon, and he was a Canadian photographer visiting Iceland to trace relatives in the northeast of the country, who had so far chosen to ignore him. He, too, was delighted to find himself "in the right place at the right time", having arrived some days before the eruption of Surtsey. Following those three short weeks in 1963, Grace's life changed. Did it stop, or did it start? I cannot say. However, she returned to our island home and, amazingly, Jon and Grace did remain great pals throughout their lives. Occasionally, they travelled together when the opportunity arose. She made only one trip back to Iceland, and as my siblings and I grew up, tales of Surtsey and Iceland grew arms and legs each time they were told.

The bedtime stories of a land over the water with volcanoes, trolls, elves, geysers, and adorable horses were for other children. Our night-time tales were of a land of ice and fire, flexing its natural muscles and vomiting steam that would burn off your skin in a moment; a place where sheep's eyes and testicles were eaten as delicacies, where infinite night skies caught your breath, and the biting cold, given half a chance, would kill you. Tales of the Blue Lagoon waiting to take a mortal soul into its rocky abyss were enough to terrify us into swearing we would never enter its milky waters. A country where the interior was so barren, to try and cross it would be suicide. Most people satisfied their curiosity by travelling there through the pages of National Geographic, from the safety of their armchairs. Our mother was so captivated by its magic that we all endured hunger for weeks to raise the airfare for her to return. Following her last journey to Iceland, she returned home with a bar of Toblerone for each of us and sat down to share her dramatic tales of adventure. No hugs, no questions as to our welfare. We children had been alone for three weeks. I was almost five years old.

Grace's future travels overseas were limited though conducted with infectious enthusiasm. She travelled only in wintertime and always to cold destinations, as though trying to recreate her Icelandic odyssey. I secretly wondered how much of her daring accounts were true, though the photographs and notebooks she amassed were proof of a kind. Either way, Grace grew more reticent and withdrawn with every passing year while Jon continued to be a dear friend and a first-class Godfather to my sister Caitlin, born two months after their first meeting.

Despite a few sporadic trips home to sober up and make amends, two further unwanted pregnancies and a particularly violent attack with an angler's priest, mother finally bolted the door on our father Dow and her abusive marriage. Though her physical scars faded, her mental scars festered ever deeper. Losing her father was a distressing blow though no tears were shed. Violent mood swings and dark depression meant we children paid dearly for our mother's lost opportunities. I came to dread the name Iceland. It felt as though it robbed us of a potentially happy life and brought with it nothing but sadness, loss, and fear.

One day when I was around ten years old, a letter arrived home. The handwriting on the envelope was no longer neat but penned by an uncontrolled hand. Our mother bit her lip as she read it. It was the first and only time I saw tears fall off her cheeks. She brought the letter up to her mouth, then placed it gently on the peat fire in the hearth. I never learned its contents. The envelope sat on the heavy oak table. Its stamp was a red and white motif of curling from the Olympic games in Montreal.

Experiences in our lives do indeed form us. Had Surtsey not erupted, had she never travelled to Iceland, Grace's life might have been different. I asked her many times why it meant so much to her. Her answer was always the same. Iceland provided her a final taste of excitement and freedom, and she spent her life thereafter knowing what she was missing. Had she not given birth to her own little "islands", as she called us children, she would have lived her life very differently. We were not wanted; we were a side product of a loveless union. To our family, to the local doctor, and to the outside world, hers was a classic case of unhealed trauma. Her low self-worth, feelings of abandonment, resistance to positive change, fear of the unknown, and the inability to move on from an abusive relationship were all classic symptoms which amplified over time. To our mother, however, it was the eruption of 1963 and the charm of Iceland that was the root cause of all her anxieties and inner demons. Life in our cold home on our grey island bored her, literally, to death.

I suppose it was little wonder then that almost 50 years later, with a suitcase of warm clothes, a tent, an old swimming costume, together with Grace's handwritten journals, I boarded the ship with my fists tight and jaw clenched. I was heading to Iceland.

BLÖNDUÓS

The definition of the word "escape" is defined by the Oxford Dictionary as "To break free from confinement or restraint, gain or regain liberty; to slip away from pursuit or peril; or to avoid capture or punishment." I didn't have to escape, and I wasn't running away. The only confinement I feared came from myself. I could "dream" or I could "do". I was the owner of my own destiny.

Since I could remember, I have written daily. Not diaries as such but accounts of places I have visited, people I have met and events that have happened around me, formulating ideas in the hopes of using them another day. Not so unusual given my upbringing. I saved money for my passport before my driving license. I booked trips to locations I couldn't pinpoint on a map. I wanted to see the world. I have followed my feet before my head. Distance and travel have been my closest friends. Many years ago in Burma, an Austrian lady I met suggested that I suffered from "No Key Syndrome". She was a member of that club too, she said. It took me some time to understand what she meant. Keys to an office, keys to a car, keys to a home, keys to doors or objects that limit one's movements, keys came with responsibilities. The idea of living without them was interesting, but in truth, I could never entirely give them up.

My coping mechanism throughout life has been to daydream, allowing my mind to travel to an unknown place I'd one day visit to embark on my own personal odyssey. Not a holiday, not forever, but for a significant amount of time. Despite working on overseas contracts and even moving to Sweden from the UK some years ago, these moves didn't count. They were real life, and with real life came real responsibilities – and keys! I started my own business renovating properties, and when there was a downturn in the property market, I found myself running a bed and breakfast. Both the business and my life became fully booked. I worked from early morning until late into the evening. I made new acquaintances quickly, and old friends from home came to visit. My decision to live abroad was not a dream. It was a calculated choice based on quality of life and opportunities. It was by no means the journey of self-knowledge I daydreamed of. Yes, the experience brought about personal growth, but there was never time for self-discovery.

The years spent overseas have been challenging; not least moving house several times, running a business and learning a new language, an impulsive marriage, and a terrifying burglary which resulted in me being drugged, robbed, and my home ransacked. The burden of it became too much for me to cope with, and I hit an invisible wall. I felt broken inside and had no one to turn to. I had worked hard for so long, doing what I thought was right, carving a life out of little and supporting everyone's emotional needs but my own. I was good at making things happen. I had even written and published my first book, and that left me longing to write more. The burglary had left me feeling violated; staying in the house where it took place was unimaginable. I learnt from that experience it wasn't always what people took from you, it was what they left you with that caused the most distress and trauma. Perhaps I wasn't so different from my mother, but I refused to settle for a life full of fear, suspicion, and hate. Through my own fog, I knew I had to find the strength to make changes, to leave behind the toxic relationship I was in and to recover from the psychological wounds the series of events had inflicted upon me. I reassured myself over and over that by journeying alone, I would eventually find a path to good health, renewed optimism and, with that, opportunities. I wasn't sure I believed it, but I knew I had to try.

To disappear for a while was therefore not utterly unrealistic, and I knew I would use my time productively writing, exploring with my camera and perhaps even learning a new hobby like knitting. I envied people who could sit on long train journeys or pass quiet evenings at home creating something beautiful and, at the same time, probably warding off the early onset of arthritis. I wished for the luxury of feeling safe. To be rid of the cancerous growth of fear that had attached itself to me. To breathe deeply and lose the vice of tension that gripped and constricted my throat. I wanted to enjoy life as I had done once upon a time. To take long walks alone in confidence, maybe stopping off at a small café for a coffee and hearing the latest local gossip. On dark, clear evenings, to lie outside on a blanket gazing up into the night sky, holding my breath in anticipation of the colourful aurora that might soon be dancing overhead without being asked, "Will you be long?" Above all, I wanted to see new places, observe, and experience everything around me and follow each day as it came. I hoped to understand what captured my mother's heart almost 50 years ago, and I knew that one day I must see Surtsey with my own eyes before it disappeared back into the watery depths and the history books.

I decided that I would base myself in a village or small town with just the necessities to hand, enough to survive, yet not too many luxuries or distractions. A backdrop of stormy seas would provide the ever-changing scenery, with a landscape of soft hills and valleys where I could wander freely, where crime figures were low, and where I didn't have to be afraid for my wellbeing. I could fish in pristine rivers, trek across pale blue glaciers streaked with liquorice, or bathe in shallow warm pools while witnessing majestic night skies. My surroundings would stimulate me and bring out a buried artistic side, or so I hoped. It felt possible anyway. When I mentioned this to my family, they scoffed and asked, "What artistic side?" What I didn't tell them was that there could be only one place to accommodate me if I stood any chance of recovering from the past years and truly be myself again. It was also time to face up to the traumas of my childhood and make sense of the psychological rollercoaster I had ridden most of my life. I couldn't ignore the emotional collapse recent events had brought upon me, nor could I go forward without dealing with the mental exhaustion ten years in a lonely marriage had left me with. I had lost sight of who I was, and I could no longer ignore the price I was paying. Despite my tumultuous upbringing, I loved my family. Mother had died alone by choice. I had little left of her other than her fountain pen and her leather-bound notebooks, and when I looked in the mirror, her image looked back at me. I couldn't visit her grave as she expressly wished not to have one. Instead her ashes were scattered in a place sacred to her. At this pivotal time in my own life, I knew that if I was to move forward, I must go back to the beginning. I needed to travel to a place I had known forever yet never visited. I needed to go to Iceland.

PLANNING

As with any great journey, it began before it got underway. The planning took over my life in the few short weeks between deciding to go and setting off. My time was spent packing up my current life, or at least putting it in storage for a while. I had no difficulty rising to a new challenge once I put my mind to it, although even by my standards, this was a large step. If my mother could go to Iceland in 1963 for 21 days, I surely could manage 21 weeks.

What would it cost, where would I stay, what should I pack, how would I even get there? So many questions to which answers had to be found. Enough of dreaming. I decided to act. It was a regular Monday evening, with an all too familiar glass of wine in hand, as I opened up my laptop and searched. A few emails later and I sat back, smiling and satisfied. The wheels were in motion. The following morning, I started to pack up the house while also waiting for the estate agent to reply to my email instructing the sale of my property. My grandmother Eliza's warning, "be careful what you wish for", leapt into my mind. I promptly bid her not to interfere, just this once.

I had decided that I would not see the *real* Iceland, meet the *real* people or indeed be inspired to be creative if I restricted myself to living in the capital, Reykjavik. I needed to be more remote. Looking at the map, Iceland resembled a haggis and looked quite small. Given that the main inhabited areas were on the coast, there didn't seem a vast chance of getting it wrong. There was a ring road running the circumference of the island, which suggested that within a few hours, I could, if I had a car, drive wherever I wished – or so I naively thought.

The planning and preparation of any project was something I enjoyed and thankfully was a task I had some experience of. I'm neither particularly brave nor naturally artistic, though I do have tenacity and resolution in spades. Surely if anywhere could bring out my arty side, it would be Iceland. Weren't they famous for their knitting and love of books? Natural creativity I might lack, but now that I had formulated a plan, I had difficulty letting it go. Thoughts of Iceland had been with me forever in one form or another, and I knew this was the right time in my life to go.

Not only do events throughout life form us, but our early upbringing also had a lot to answer for. Growing up on an island meant most of our clothes were handmade. Being the youngest of three children, I had hand-me-downs until well into my teens. I had destroyed the earliest photographs that showed me in an array of knitted swimsuits, trousers, and ponchos two sizes too big, much to the anger of my sisters, who found them hilarious. I couldn't wait to leave home and earn enough money to buy my own real clothes. On packing for the trip, I realised almost nothing I owned was made of wool. One cashmere jersey bought in Ireland in a moment of retail madness, a Pashmina scarf from a dear friend in Kashmir and an original Rina Da Prato cobweb shawl gifted to me by the artist herself. Wool reminded me of my childhood.

Plans got underway. The estate agent had replied, and the house would be on the market in days. The following Thursday, I received my first email from Iceland, offering me the opportunity to stay in a small town, with an active community, famous for its textile connections, with the added opportunity of knitting lessons, should I wish. I was giddy with excitement and readily accepted. Later that same day, I opened the email attachments to read further about the living arrangements, what it would cost and information on how to get there. I wouldn't be alone, it assured me, as the accommodation would be shared with a group of international textile artists. I hadn't a clue what a textile artist was. I imagined tapestry, embroidery, and weaving. A cluster of vegetarians sitting around their spinning wheels and looms wearing rainbow-coloured clothing. I don't know why I thought that, but I wasn't so far off the mark.

Back in my childhood home, there was always clutter from ongoing projects. Knitting wool and tapestry threads stuck to every sock, and pins found the delicate soles of my feet daily. If I complained and tried to tidy away the clutter, I was threatened with being sent to Surtsey. I hated it, yet here I was almost four decades later being offered an opportunity to live in a house in Iceland, full of artists, for up to six months and accepting it! I scalded myself for being so judgemental. After all, how difficult could it be to share with some nice artistically inclined individuals? They might even prove helpful in teaching me a skill or two. Anyway, the accommodation was a backup. I would take a tent and sleep outside as often as I could, though I knew I needed to shower and wash clothes, cook food, and charge my electronics each week. I also wasn't sure just how cold it got in Iceland in winter. I had already emailed my acceptance and bank details, so there was no turning back. I was off to Iceland. To be precise, I was off to Blönduós.

"Where the fuck is Blönduós?" my best friend Mats asked, rather startled when I called him to explain what I planned to do. He was my oldest friend from childhood and my dependable hero. A retired special forces operator, he now volunteered with mountain rescue and sold insurance. He knew a thing or two about survival.

"It's north of Reykjavik," I replied confidently.

"Everywhere is north of Reykjavik, Emma!" replied Mats. "Where exactly?"

As we spoke, I quickly opened the laptop and searched for a map. Eventually, I found Blönduós.

"Very north," I muttered. "Eleven o'clock on the map."

When I called my eldest sister later that same evening to share the news, she immediately asked, "How do you spell Blönduós? I can't find it."

I wasn't certain myself.

"It must be near the Arctic Circle," she said.

I started to worry.

"Hope it's not near bloody Surtsey," she joked.

Eventually, the jokes faded, and seriousness set in. The more I spoke of my plan, the more enthusiastic people became. Even my nieces and nephews, who rarely travel, thought it would be cool to visit Iceland for Christmas. So much for being alone, I sighed, though deep down, I knew it was unlikely they would come.

I researched Blönduós once again, a little more seriously this time, and was stunned. There appeared to be little apart from a few houses, many dilapidated buildings, and a slaughterhouse. The town was divided in two by a river. Surely this couldn't be the town in which I had agreed to spend an entire winter. The only positive search I found advised me that Eric Clapton fishes here annually. I guessed he might be more elusive than the Northern Lights, but I felt there must have been at least a decent pub there, if not for Eric personally, for his entourage, who I assumed would follow in his wake. Maybe Blönduós wasn't so bad after all. Additional information came through from the artist residency by email. They attached a Liability Waiver reminding me that legal action of any kind was pointless, a copy of their alcohol and drugs policy, acceptance of my initial request form (it was a residency and not a hotel, you had to apply and hope you were accepted), and a brief introduction to the town. It certainly didn't mention a pub. In fact, it read something along the lines of "Blönduós is the most populated area of Húnaflói bay in Northwest Iceland. Around 850 people live here permanently, with a transient population nearing another 150 or so working in the factories. It is a pit stop for those travelling the Ring Road and serves as a service point for local farms. It has a large slaughterhouse and a dairy, a wool washing plant, and other light industrial outlets. Accommodation is basic but plentiful. There is a café, a petrol station, a swimming pool, several museums, and a large hospital with a well-appointed unit for diabetes." It went on to say they'd had "moody weather of late, so bring a hat that fully covers your ears, and a sturdy pair of galoshes to be worn with two pairs of socks, the outer one being wool!"

I would need a car then. The residency would be my base where I could store my possessions and sleep when I wasn't camping. I planned to explore as much of the country as possible, but it felt safe to have a place from which to work. Despite a few initial concerns, I continued with my plans. As far as Google Earth was concerned, Blönduós was the arse bone of nowhere. Trip Advisor was even less kind, but I was determined to remain positive.

Although Blönduós wasn't Reykjavik, I was sure that if 850 people lived there and visitors travelled there from overseas, they couldn't all be wrong, could they? Although it did seem rather a small town, it would be near larger towns. Akureyri wasn't far off to the right of the map, and it had an airport. Iceland really wasn't such a big country, and I could drive wherever I wished if I felt like a change of scenery. Didn't the best writers and artists escape to rural retreats to find themselves?

My stubborn streak didn't let me down, and as I had already paid in full and told everyone that I was going, I had to see it through. By the time I plucked up the courage to share the news with the remainder of my friends and family, I had embellished it to sound as though I was escaping along an icy version of the yellow brick road to a fairy tale land. How lucky I was to be bunking off from the daily grind and heading towards smouldering volcanoes, blue lagoons with the odd famous Hollywood star thrown in for good measure. Weren't they always filming the latest movie or music videos in Iceland? As I reeled through my mantra, which included casting alongside Eric the angler Clapton and toe-touching Bear Grylls in a hot pot, I couldn't help but be a little disappointed by the same repetitive questions.

"Why Blönduós and where the hell is it anyway?"

I wondered why it was so important where it was. Couldn't it just be somewhere, anywhere?

Time was marching on, and I aimed to be as productive as possible within the short time I had left to organise the finer points of my journey. What should I pack, how cold could it get, where would I receive post *(what post...?)*, and more importantly, how was I going to get there? By now, I was set on taking my own car. As I searched for useful apps to download, including Icelandic traffic signs, rules of the road, and ICESAR, the Icelandic Search and Rescue service, I was glad I had been living in a Scandinavian Country for a decade and therefore was not wholly unprepared for what lay ahead. I understood that if I took my own car with me, I would also need to take my winter tyres; Iceland appeared to have the same winter vehicle requirements as Sweden. They would take up space in the car, so I'd have limited room for anything else. I toyed with how this would work en route to the only ferry, which operated from Denmark, and where it was prohibited to drive with studded tyres in winter. The ferry left from Hirtshals in northern Denmark, and there was no way of avoiding driving there. The ferry from Gothenburg in Sweden would cut the journey down dramatically, but I would still have to drive for around one hour through Denmark and wasn't convinced the police would be lenient with me if I got caught breaking their laws. My other quandary was that if I drove out with summer tyres on the car, how would the roads be once I arrived in Iceland? Would it already be winter there? I had looked at a map and discovered that the ferry arrived in a small town on the East coast, which appeared to be at the foot of a very steep mountain pass (more on that later!). I decided to telephone a garage in the area and ask their advice. Despite numerous attempts, they didn't answer. I called the tourist information at the ferry terminal, and the conversation went something like this:

"Hello, is it alright if I speak English?"

Silence.

"Hello," I asked questioningly, not sure if they had heard me.

"Yes?" a male voice replied.

"Oh, hello," I said for a third time, a little thrown. "I'm calling from Sweden and need some advice about whether you may have snow in October."

Silence again.

"Hello."

"Yes," he replied again.

"Oh, you are there." I ploughed on, explaining my predicament, unsure whether he had understood what I was trying to say.

After an achingly long pause, he replied in English, accented with a broad dialect that was hard to understand. Sadly, he said he could not guarantee snow in October but there was a good chance of it. He cautioned that I would need winter tyres for the winter months, and he also confirmed that, as far as he understood, I could not drive with them in Denmark. The best advice he could give was to call a garage. I was no further forward. My stubborn streak proceeded, and I pressed on. I explained I had, in fact, been calling the garage in town, but they never answered the telephone. He told me the proprietor of that garage was quite deaf so probably didn't hear the phone. He suggested I contact another garage, some 21 kilometres away in the town of Egilsstaðir. He talked in his faltering "Icelish" about the difficulties of driving the mountain pass in ice or snow with summer tyres and suddenly raised his voice and went on to tell me it wasn't wise, or perhaps even possible.

"Suicide," he shouted down the line. "Suicide. Lady, do you hear me?" he shouted down the phone to make sure I understood.

"Yes, I understand," I stammered.

Bloody hell, I thought. I hated driving in ice, and the mention of the mountain pass had certainly got his attention! I ended the call by saying I would keep ringing the local garage to book a time on arrival to change my tyres. Now he was on a roll.

"No, that wouldn't be a good idea."

"Why not?" I asked.

"It isn't a good idea at all. Don't do it."

"Why ever not?"

"It isn't how it's done."

"Then how is it done?" I pleaded.

"You don't need a garage. We will find someone to help you. Maybe I will do it."

"Will you be there when I arrive?"

"Yes."

"I haven't told you the dates yet?"

"I'm usually here when the ferry comes in."

"And you will help me change my car tyres?"

"Maybe. Come and find me," he answered and quickly hung up.

My plan of leaving on 1st October seemed to be causing such unnecessary difficulty with the car that I contacted the residency to enquire whether I could come earlier to Blönduós. They replied that they could accommodate me only if I could arrive on 1st September. This would, I decided, be much more practicable and probably a heck of a lot safer too. My plans were brought forward, and the winter tyre dilemma eased. I could arrive in Iceland before the real winter weather took hold and then arrange for the local garage in Blönduós to change my car tyres when it was time. Surely the first days of September would be clear of ice, and the mountain pass from the east coast towards Akureyri and northwest to Blönduós would be glorious in its autumn colours. I was barely sleeping following the break-in, and leaving a month earlier would be a blessing. I couldn't wait.

Plans were fast-forwarded. Even my most cautious of friends were caught up in the final flurry of excitement.

Their only sadness being they couldn't accompany me on this adventure.

Mats and I chatted almost daily as he was curious how the final stages were panning out and demanded I forward him my packing list. He enquired what safety measures I had considered; had I packed a spare shovel, rope, spare batteries (for everything), an extra daypack in case one got wet or the straps broke, and of course, layers of practical clothing. The seven Ps were never far from his mind, and he kept checking I understood them too. Proper planning and preparation prevent piss poor performance.

"You're not twenty years old anymore, Emma," he reminded me.

Hurtful but he meant it kindly, and he was right. I wasn't far off 50, with a history of back surgery, asthma and predisposed to feeling the cold, I guessed I should take my piss-poor planning a bit more seriously. I didn't mention to him that I suspected I was also entering perimenopause. Maybe Iceland would ease my burning hot flushes!

My list of must-have items grew by the day, and when I was eventually ready to seal my boxes, Mats flew over to Sweden to take a final look at what I had packed. He checked everything thoroughly, threw out a few unnecessary items then added quite a few more essentials. He accompanied me to Gothenburg, and we went straight into Naturkompaniet, a favourite store among outdoorsy types, that saw me trying on an array of very unflattering items, from baggy trousers to string vests and Long-Johns, the ultimate in warmth versus breathability, or so the marketing would have you believe. I wasn't convinced. They looked to me like wool.

Anyway, if I was going to die from hypothermia in the frozen north, I wanted my final moments of being found to be that of pity, not ridicule. Frozen to death and found some time later wearing an extremely unflattering string vest and matching trousers was not in my plans, even if they were of the finest merino wool. I would forfeit a small amount of comfort for an ounce of dignity. We agreed on some Icebreaker merino wool base layers in plain black, which turned out to be my best investment in years.

Having spent many a night in the cold and wet in the military, Mats assured me I would be grateful for these extra items. I didn't wish to dampen his enthusiasm or support by telling him I'd probably skip the camping and find a hotel. What if I got stuck for hours and perished in my car or fell down a hillside and twisted my ankle, to be found days later by search and rescue wearing inappropriate footwear. If the cold didn't kill me, maybe the rescue team would for being so stupid. What if frostbite caught me out while I was watching the northern lights and I had to have bits of me amputated. Could I forgive myself if I hadn't been wearing the Lundhags, Fjällraven, North face and Icebreaker lifesavers I was now investing in? I use the word investment by choice. The cost of this trip was mounting up.

As we went through the finer points of daypacks with the sales assistant, I was both impressed and bewildered by the choices. We spent an entire afternoon in the store trying out different models, sizes, and weights.

"Are you joking?" I gasped when he told us the price.

"It does have side pouches and is water resistant," chirped the salesman.

I liked the bright yellow pack, which I expertly explained would be easier to be seen by search and rescue. However, both men replied that in such a situation, red was best. Also, Mats reminded me yellow attracted wasps and went on to explain to the salesman that I was highly allergic. "There can't be so many wasps in Iceland in winter", he sarcastically replied. He was right, but in Faroe on the way there, it was swarming with the things. I went off and busied myself looking at foldable cutlery (spork!) and coffee makers. I looked at the prices and thought, bugger that, I had a car so the least I could pack was real cutlery and my Illy coffee maker. I did buy two packs of Lemmel camping coffee, however, for authenticity.

By the time we made it to the tents section, I'd almost had enough. "Aren't all tents made for outdoor use?" I retorted impatiently when Mats and the salesman discussed materials.

Mats remained cool and calm and pointed out the pluses and minuses of all this information with ease. The shop assistant loved him. He was therefore sad to hear it was only myself that was going on the expedition. I had a little fun tormenting them both with childish comments, having already decided I would buy whatever Mats recommended. A tent in winter really was life or death – even I knew that.

Given my budget and under-enthusiasm for overspending on camping gear, I settled on the essentials only, which were the tent, Therm-a-Rest, sleeping bag, a day pack, and of course, the coffee. I already had a small Primus burner and storm kitchen which I occasionally used at weekends. I also bought Icebreaker merino wool thermal trousers and jersey base-layers, a Fjällraven windproof jacket, Lundhags polar quest boots together with a tin of their own label grease, a pair of Hestra leather gloves with separate liner, and six pairs of Woolpower socks. On folding the jacket at the check-out, the sales assistant asked whether I wanted some Greenland Wax. Apparently, the jacket wasn't waterproof. He explained I would need to treat it with wax. Coming from Örnsköldsvik in Northern Sweden, you would think Fjällraven would have an inkling that I might wish to use it outside, and, therefore, it would be very helpful for it to be rain resistant at the very least, but apparently not. I bought the bar of Greenland wax, with instructions on how to iron it on. The salesman held a particularly long smile in my direction as he reminded me, "Just three easy steps, heat, apply, repeat!" He knew as well as I did that I wouldn't use it.

By the time my own personal D-Day arrived, I was suited, booted, packed, and hadn't felt as excited in years. My electronics were fully charged, batteries and spares checked, and apps uploaded, including a handy 112 (the equivalent of 999 in the UK) just in case I got into serious difficulties. I was certain some of the kit I had accumulated was a little overkill, like the mountaineering gloves worthy of summiting Everest on any day of the week. The Lundhags Polar Quest Boots, with their wool felt inner liner and traction soles, were also sure to go unworn as their natural element was surely a Siberian mine (they were in fact my biggest disappointment as they gave me no traction on the ice). It was Iceland I was going to, I kept reminding Mats, not the Arctic or Russia. I hadn't told him that if I survived this trip, I was already considering future trips north including Qaanaaq.

THE JOURNEY

The alarm rang at 04:00 hours on 28th August. I was already awake, having dreamt of lying alone in a cramped tent in the interior of Iceland, near a lake and with twinkling stars above me. A huge polar bear was impatiently circling my tent, having upturned the remains of last night's dinner, and knew there was more to be had from inside the tent. I had no protection other than a flimsy bit of nylon and my sleeping bag; in other words, I had no protection at all. The half-ton hungry bear would rip me to shreds, piece by piece. Could it smell how much body fat I had? Probably. As with all my dreams, it was so vivid and real. Sleep, for me, was rarely a safe place. The bear's putrid, nasal breath heaved while the slap of its huge feet shook the earth as it thudded and grunted outside the tent. I was burning hot and hyperventilating with fear, shaking with sheer panic. At least I wasn't drowning.

I instantly went off the idea of camping. The fact that there aren't any polar bears in Iceland didn't help my nerves. It wasn't until much later that I learned they do, in fact, on occasion, have the odd polar bear come ashore. Worse still, the tiny town of Blönduós houses the only dedicated polar bear museum in Iceland, with a taxidermy specimen of a tragic hungry female who did wander ashore some years ago. She wasn't the only polar bear to have reached Iceland. In six decades, there have been at least a handful of polar bears who have made it to Iceland, enough for the government to have official legislation about how to deal with them when they did swim ashore. No bear has killed a human being there – but there must be a first, right? Perhaps a hungry bear would dispose of the evidence, and I would become the unsolved case of the abandoned tent. Not quite the grand exit I had thought of for myself. My dream had scared me, and despite waking up and realising it was just that, a dream, or rather a nightmare, I started to think what I would have done to fight back if I did meet a bear. There wouldn't even be a tree to climb. Bears were excellent climbers anyway, far better than me, I imagined, as there were no trees to climb where I grew up. Pity the helpful shop assistant in Gothenburg hadn't recommended bear repellent. I'd have been far more interested in buying that.

A little unnerved, I got up, packed the final items, and plodded through to the empty kitchen for the last time. The reality of what I was now doing dawned on me. This would be the last time I would wake here, look out over the garden and pop outdoors to fill the bird feeder. I hoped the new owners would have more luck in the house than I had.

As I loaded my final luggage into the car, including my favourite memory foam pillow, a set of crockery and cutlery for two (ever optimistic), some food staples and one or two homely comforts to make the winter cosy, I made one last tour of the house. Had I forgotten anything? I grabbed the heavy-duty torch and battery charger from the wardrobe under the stairs, understanding it would be pitch black in winter, and I was not completely comfortable with the dark. The week previously, I had also invested in a head torch that I could charge through a USB cable. Both my computers were packed along with a rather expensive camera and three lenses which I had wished to buy for years. This trip had been the perfect excuse to bite the bullet and splash out. I looked around a final time and said farewell to the house.

Dressed in shorts, a summer shirt, and sandals, I picked up the house keys, locked the front door and posted them through the letterbox. I knew I had over-packed and would never use half of what I had with me, but it felt comfortable to take a little of home along too. The lock-up I had rented was packed to the gunwales, and I planned not to open it up until I had returned and found a new home. I wouldn't think too much about that now as who knew where my adventures might lead me. Driving out of the front gates for the last time, I patted the seat where my handbag and camera lay and headed south on the E-6 motorway to Gothenburg to catch the first of two ferries to Iceland. Finally, I was on my way.

From home, it took barely an hour to reach the Port in Gothenburg, where the Stena Jutlandica left for Fredrikshavn in Denmark. I boarded through the vehicle lane, parked, and made my way upstairs to take a seat in the restaurant, and drank an alcohol-free toast to new horizons. My constant mental battle with "what if" clouded the moment a little, which it often did. I should have been like a child on Christmas morning, wide blue eyes blazing in delight at the adventure ahead. Instead, I was over-analysing every moment. The crossing to Denmark was a little over three hours and the weather was perfect. Arriving on time, having enjoyed an impressive buffet lunch on board, I drove off the ferry and followed the road signs to Hirtshals, arriving just as the sun was setting over the harbour. As I parked the car, I worried about leaving it fully loaded in the quiet, unlit car park. I moved it nearer to the only available streetlight, locked it and prayed nothing untoward would happen. I felt my life was in that car. I knew I wouldn't sleep much as tomorrow was a big day.

Daylight broke and I couldn't wait to explore the town. Hirtshals was situated at the top of the Jutland peninsula in the north of Denmark. By the look of it, the harbour was its main attraction. The Port services many ferries, including those to Norway, the Faroe Islands, and Iceland. There was a large aquarium, the North Sea Oceanarium, which was apparently one of the largest in Europe. There was a decent number of visitors, given that it was late summer and generally poor weather for the time of year.

I found a pretty café on the seafront, not too far from the ferry terminal and from where I could park the car in full view. I opted to sit outdoors and watch the local anglers cast their lines over the pier. I ordered a large hot chocolate and waffle, which came served with cream and jam: I was hoping the sugary explosion into my blood might give me back some energy to see me through the day.

I felt the first pangs of real nervousness as I paid the bill and walked back to the car. Procrastination over; it was time to make my way to the ferry. M/V Norröna was leaving in less than three hours, and I had been advised to be in the queue no later than two hours beforehand. As I drove the final metres through the entry gates, I checked for the hundredth time that my purse and passport, keys and bankcards were all in order. I found I did this a lot, checking and over-checking that my handbag was safe. I don't think of myself as compulsively obsessive, just that if I did lose something fundamental when travelling, it would be a huge inconvenience. Most of my life, I have lived alone and without anyone to call upon to help me out of sticky situations. Better to just be careful and check along the way that everything was as it should be. My family, on the other hand, teased me that I had every obsessive disorder under the sun.

The Norröna, berthed in the small harbour, looked majestic in the midday sunshine. The flagship of the Faroese Smyril Line, operating year-round between Hirtshals and Seyðisfjörður, offered me not only the opportunity of taking my car with me but also, hopefully, an enjoyable sea crossing. The heavy traffic of the summer season had subsided, so now the ship should be a little quieter as I hoped to relax after the eventful last few weeks. The crossing took three days and nights so there would be time to enjoy the facilities on board while keeping an eye out for humpbacks, orca, and even blue whales if I was lucky enough. I silently prayed for good weather. As I headed for the boarding queue, I managed to nervously laugh out loud. I was surrounded by a worrying array of badass super jeeps, the customised type that had every kind of warning system, lighting, and the most enormous tyres I had ever seen not attached to a tractor. I learned later that Iceland had arguably the best modified-truck scene in the world. Almost every car looked as though it was ready for battle. People were running around, emptying hard cases of equipment, dressed in colourful heavy-duty outdoor clothing that made my shorts and sandals look silly. Had I misjudged the weather in Iceland so much? Most number plates I observed were either German or Icelandic. Perhaps they were planning trips into the interior. Suicide, I could hear my mother shouting. I've no idea where the actual "interior" was, though I imagined it was somewhere in the middle of the country. I hoped it wasn't near Blönduós! There was no backing out now. I had come too far with my plans and would see it through. Even I couldn't back out before I arrived!

The Norröna was purpose-built in Lübeck, Germany, in 2003 by Smyril Line to service freight, passengers, and vehicles between Denmark, the Faroe Islands, and Iceland. From the look of her in the harbour, she was up to the job. Part freight vessel, part cruise ship, I looked forward to exploring the bars, restaurants, and shops, not to mention the hot tubs on the upper deck. As I had booked the crossing rather late, the more luxurious cabins were taken, but I was more than happy with a standard cabin with bunk beds and a small window, offering adequate space for the short cruise. The sailing time itself takes around 45 hours in the summertime, though in winter could take as many as 53 hours, not including layover time in Tórshavn. I tried not to think of the return journey, which would be in the middle of winter. While boarding the ship, I noticed the lifeboats and knew these were self-righting. This was one consolation that came about following investigations into the Longhope Lifeboat disaster.

This wasn't my first trip with Smyril Line. Thirty years previously, I had taken the original MV Smyril from Aberdeen to Lerwick (Shetland). She stood in for MV St

Clair, which was having a refit. It was a winter trip, and the weather during the entire 18-hour crossing was dreadful. The ship tossed, rolled, and heaved, as did the entire passenger list and most of the crew. One hardy sailor I forever remember was Eric, with a red beard and bushy hair, who noticed me sitting alone in the cafeteria. I had booked only a couchette but couldn't stand the stench of people vomiting, so decamped to the area where no one else was, the cafeteria. When the crew cleared away the sandwiches due to lack of customers and retired for the evening, Eric came to chat me up. He tried various tactics to cajole me into going on a tour of the staff quarters. No thanks, I firmly replied, as I gagged from the stench of stale alcohol on his breath, though found myself over-politely asking: was it possible to order some tea and toast? To this day, I am ruled by mealtimes and hate being hungry. In fairness, he did, without much in the way of customer courtesy, show me the galley and told me to help myself. The trip had been one long adventure and when we eventually made it into Lerwick harbour, the gangplank blew off its couplings, and we waited several more hours before we could disembark. I hoped three decades had seen changes for the better in Smyril Line's staff training.

My nerves subsided as I climbed the companionway leading from the car deck and retrieved the card which I had been handed with my cabin key, finding my berth on the 6th floor. The cabin was spacious and comfortable. It had a tiny loo and shower, spotlessly clean and made up with fresh towels and soap, a TV and even a small wardrobe. It was precisely as the brochure had described it, all you needed for three nights on board. I headed off to find the bar.

As the ship set sail, the late afternoon sun bathed the bay of Hirtshals a golden orange, and I silently bid farewell to Denmark for a few months. This service was a lifeline for the Islands and yet I had never met anyone who had made the journey. Most people I knew from the UK rarely took their car on holiday unless they were going to

France or possibly Italy. Why would they if they could fly? Likewise, in Sweden, people simply didn't drive to Iceland; they took a flight for a short break and returned home.

With a large Tanqueray and tonic in hand, I headed out to the sun deck to enjoy the view. It was deserted and I wondered where everyone was. I guessed they were packing and repacking their huge expedition holdalls. My mobile phone beeped with a message from Mats. He had uploaded a marine tracking app and could see I had set sail. He proposed a toast to a girl with wonderfully strange ideas and signed it, as per usual, with 'She Who Dares, Wins' and – not so usual – had added two love hearts. With my dearest friend holding an eye on me, I no longer felt alone. It was a comfort knowing Mats was watching over the ship. Not for the first time, I questioned why I had not married him. We had talked about it. Our backgrounds were similar. His desire to travel and his love of the outdoors took him into the armed services, and as with everything, Mats excelled and flourished. He never married. When, eventually, I did settle down, I remembered the phone call well. I was in Fiji on holiday with a man I'd very recently met, who unexpectedly proposed after only a handful of dates. I rang Mats to share my news, realising sometime afterwards I was secretly asking him what I should do. He said little, wished us both the very best and declined an invitation to our wedding. His only words were that he hoped the new man in my life would not turn out to be a rat. At that moment in time, sailing out of my comfort zone, knowing Mats was with me, albeit through a Marine Tracking app, I felt happier than I had in years. We sailed out into the Skagerrak, the lighthouse fading into the distance, and I made silent bets with myself of how long it would be before I would spot a whale. What I didn't yet know was that it was to be the last message I would receive from Mats. Duty called him unexpectedly out of retirement and an IED claimed his life in Afghanistan. I received the news a short while after arriving in Iceland.

Watching the clock for the restaurant to open, I was there early to ensure a window table. I stood and waited, yet nothing happened. I took another tour of the deck and then returned to find the restaurant still closed. Frustrated, I marched to the information desk and asked: was there a problem? To which, the purser explained that all the times quoted on the ship were "ships time". That meant that as it was a Faroese Ship, they ran on Faroese time. I muttered an apology and felt foolish as I walked away. Why hadn't I remembered that! Six o'clock on the dot Faroese time, the restaurants opened, and I chose the steak house. It was the à la carte option, though I wasn't too convinced it would be anything special. I was mistaken. I ate the most delicious, succulent steak I had had in years, paired with a superb glass of Bourgogne. The friendly staff, together with the alcohol, helped me to relax and I started to enjoy the cruise. I took a final lap around the decks and retired to my cosy cabin. If this was the new Smyril, I was impressed. In fact, I was already considering future trips.

As night approached, the wind increased a little, and I hoped the waves would not grow in intensity. I had never been seasick, but my fear of drowning was never far away. I brushed my teeth, crawled into the small bed, and blissfully slept like a log. The following morning, we were completely at sea. There was no land visible, and only the occasional large tanker inched by on the horizon. I unpacked my new boots and the boot grease I'd bought and settled into a deck chair on the upper deck in the sunshine, busying myself waterproofing my new heavy Lundhags. It was fresh and a little windy but still pleasant while the sun shone on my face. As the morning passed, I enjoyed a leisurely brunch and then returned to my favourite spot out on deck. *Why hadn't I done something like this years ago*, I thought. I read the newspapers from days before, checked emails and settled into life on board.

The sea remained calm, and the gentle rolling of the ship was a comfort. It was far more luxurious than I had imagined. I woke myself up snoring lightly on more than one occasion. Onwards the Norröna motored northwards. The journey was spectacular, especially on approach to the eighteen or so islands making up the Faroes. Late morning saw Eivin, a local politician, hoisting the flag over Tórshavn as our ship approached. Grabbing a light jacket, I wandered off, camera in hand, to enjoy the short shore leave. A friendly chat ensued with Eivin, with him passing on the information of the upcoming vote for the 38 elected Parliamentarians the following day. Enquiring what were the main topics and issues in the North Atlantic, his answer was surprising. Not fishing quotas nor wind power, no longer the difficulties facing the Islanders due to migration or even the building of impressive tunnels and bridges the islands have seen over the previous years. The (flagpole) topic on everyone's lips was Gay Pride. The Church was at loggerheads with the Youth, he explained, the elders voting for solid Spirituality and its defined behaviours. The youth were more interested in same-sex rights. I quipped they'd next have a Gay Pride Festival and he advised me that was last week. As I wandered through the streets of Tórshavn, colourful remnants of chalk rainbows were evident on the roads and pavements, confirming this.

Wool and postcards were at the top of my shopping list and, of course, knitting needles. I'd never owned a pair! After several requests for directions, finally made sense of by the local traffic warden, a beautiful bright-eyed and clear-skinned girl, I eventually arrived at the wool store. Navia was the busiest knitting wool shop or garnaffär in town, a haven for both men and women, fingering the pages of the latest knitting patterns. I guffawed a little. Hadn't these gone out of fashion in the 1970s? Certainly not on Faroe by the looks of it! A full hour later saw me leave the shop armed with what the assistant had recommended. Wandering back towards the old harbour, I stopped by the local fishermen selling their catch. Surprisingly this was not fish, but birds, Fulmars to be precise. Suffocated, plucked, and cleaned, ready for poaching or roasting. I immediately wracked my brains for a suitable recipe but found myself wanting. I would have to ask for cooking tips. In attempting to do so, the conversation too swiftly turned to wool. All three fishermen wore the traditional Skipstroyggjan or Faroese Boatman's jersey. Currency for many years, these Islanders certainly knew what to do with wool. All three jerseys were well worn and loved, fitted well, were striking in pattern, not garish like many Scandinavian patterns but natural and homely. *Like the people*, I thought. They spotted the familiar logo of the local wool store, snatched my carrier bag and inspected its treasures. Perfect, they announced in unison. You chose the best wool on the Island and the right size needles too. Where else in the world could you stop three young men beside a harbour and have them offer you knitting advice? In my enjoyment of the moment, I forgot to ask for an appropriate recipe for Fulmar, but it was always good to have a reason to return to a beautiful place. On saying goodbye, I took a quick photo and headed back to the ship with minutes to spare before we were underway once again. As we left the harbour, I was rewarded with two beluga whales cutting a wake in the late afternoon sunshine and, looking around to see if anyone else had seen them, received a thumbs up from the crew who, too, had spotted them from the bridge. The remaining sailing time to Iceland was around 19 hours. We were more than halfway there. I enjoyed an early dinner and drink at the bar, then, making the most of the late summer evening sun, took a dip in the hot tub on the upper 7th deck with three Icelanders heading home after working in Faroe. The hot salted water felt wonderful on my skin, and as we enjoyed a beer, a whale's fluke delighted us all. Sadly, the whaling vessel wasn't far off in the distance, and we hoped the whale would dive deeply while it had the chance.

The noise from the engines slowing down woke me. It was 06.10 on 1st September, and I peered out of the small cabin window. I hoped to catch my first glimpse of land but saw nothing: the fog was too thick to see even the waves below. However, by the time I had dressed and headed for the buffet breakfast, the weather had opened into a glorious sunny morning. M/V Norröna nudged her way through the long, deep fjord that twisted and turned from its mouth and into the shallow sound on the east coast of Iceland and into the small port of Seyðisfjörður.

I ate quickly and took my second mug of coffee out on deck. The morning breeze could not be described as fresh - it was bone-chillingly cold - though I persevered and was rewarded with unbelievable photo opportunities.

I won my personal bet with myself, having spotted several white-beaked dolphins and a small pod of orca whales before nine o'clock. Excitement took over me and I shouted "Whale!" before running to the railings with my heavy camera. The remaining passengers who hadn't given up and gone inside to the warmth were delighted that I had given them warning, and rushed like a wave to where I stood to also take pictures.

It was surprisingly windy as we found our way forward, and I seriously wondered: had I brought enough clothes with me? I was thankful for Mats' help in persuading me to buy thermals. I'd write him a postcard as soon as I could. Maybe wool was the way to go, after all. Despite the sunshine, I had dressed in a base layer plus my fleece jacket and coat, hat, scarf and gloves, and two pairs of socks with boots, yet this was no match for the wind chill. I was freezing cold, and my extremities were numb and starting to sting. It was still technically summer.

Mary and John, a couple well into retirement age to whom I'd chatted during the trip, were dressed in lightweight rainwear and looked frozen. While boot polishing, I had struck up a conversation with John, and he, in turn, spoke of their journey from the lakes of Minnesota, onto a cargo boat bound for Italy, with a Polish captain and no other passengers. They had left home seven weeks ago and were now missing it dreadfully. They were used to the cold weather, had lived in Alaska, and yet agreed this wind was something else. They were interested in my planned trip and said over and over how they envied my courage. They had researched Iceland well and seemed familiar with the towns and villages they would see through a bus window on their way to Reykjavik. They'd hoped for more time to explore but had suffered several delays and now simply hoped to catch their flight home. Several times John asked where exactly I would be based. When I said Blönduós, he shook his head and said, "No shit, where the fuck you say Blönduós was?" in his distinctive guttural articulation. An hour later, we would chat, and again, he would ask me what the name of the town was. Again, I would tell him, and he laughed and said it sounded like a disease a cat would get. I helped him find it on the map. He was sure I could have found a better place to spend the winter. I wondered if he was right.

They had spotted only one whale on their entire journey and were sorely disappointed. They thought they'd see much more wildlife. During the crossing, I had spotted some larger whales, yet each time it happened, Mary and John were on the opposite side of the ship. Finally, as I screamed "Whale!" they jumped to attention and saw the tail end of a glorious sighting, a family of orca. They were delighted. The last I saw of Mary and John, they were heading in the direction of the garage, very likely the same garage that never answered their telephone, where they hoped the bus left from. However, on which days of the week that occurred, they had no idea. They had only a couple of days to get to Reykjavik to catch their flight home. I wished them luck and hoped they made it. Sadly, my Volvo was full to the brim with no space for their luggage, let alone the pair of them and Reykjavik was in the opposite direction to where I was heading.

Seyðisfjörður, with its well-preserved wooden buildings, was a pretty town with a little under 700 inhabitants. The town could trace its origins back to foreign trading in the mid-19th century, mainly with Danish merchants. Its fortune was made, however, when the Icelandic Herring fleets, or Silver Darlings as they were known, were established by the Norwegians in the late 1800s. During WWII, Seyðisfjörður was a base for British and American forces. Sheltering and guarding the town were two imposing mountains, Mt Strandartindur and Mt Bjolfur. These steep-sided valleys were prone to avalanches, which were a risk to life in various areas of Iceland. In 1885, it was reported that 24 people lost their lives, while as recently as 1996, a factory was razed to the ground. Near the church, there was an avalanche monument made from white-painted twisted girders, some remains of the factory. In 1995, on the West Fjords, the towns of Súðavík and Flateyri were hit by an avalanche which claimed the lives of 36 people. There were 20 survivors among them. One could only imagine the scar that leaves on a small community just as the lifeboat disaster had left its scar on ours. Stories abound of heroic rescues, and indeed, the rescue services here were heroes, as I would later on learn. On that day, 600 rescue workers from the Icelandic volunteer service ICESAR reacted quickly and did an unbelievable job rescuing those they could possibly save, including an 11-year-old girl who was buried for 11 hours. Just like the RNLI at home, normal citizens offering themselves as unpaid volunteers provided the service.

I didn't have time to see more of Seyðisfjörður, as I was anxious to be on my way. As I would be returning by the same route in a few months' time, I decided that I'd take an extra night or two then to see the sights before the ship set sail homeward bound.

Having driven off the ferry, I halted for the approaching customs officer. Once again, I reached for my handbag to check it was there. I patted the steering wheel of the car. *Welcome to Iceland*, I thought, opening my window. The efficient officer demanded "Paperwork."

"What paperwork?" I muttered.

"No problem." The official sighed, obviously used to visitors coming unprepared, and took a bunch of forms from a file.

"Fill in these to register your vehicle in Iceland for the duration of your stay."

"Oh, yes, of course," I said politely.

"You sell Volvo, you pay high taxes." She winked and nodded sternly.

I hadn't planned to sell the car, but swiftly completed the paperwork and handed it back. That done, I was waved on with a "Have a nice day" in Icelandic which sounded something like *egg du goodun' dag*. Stopping for a quick photo of the Blue Church on the grounds that it was pretty and just the colour blue I hoped to knit a pair of socks in, I waved a final goodbye to the passengers whom I had met during the voyage. I let the monster trucks take the lead. They seemed as keen as I was to get their adventure truly underway.

This small town reminded me in many ways of the Scottish Isles. As kids, we played on the icy beaches or on the cliff tops with little thought to the wind or cold. Hats blew off and gloves got in the way. Despite being dressed in several layers and the car heater turned up a little higher than usual, I passed two young boys dressed in simple rain jackets and no gloves playing on the water's edge; one had a fishing rod while the other whistled to the seals. They were laughing and looked happy, completely oblivious of the weather, despite their hands and faces purple with cold.

THE INTERIOR!

Tears fall in all the rivers: again, some driver pulls on his gloves and in a blinding snowstorm starts upon a fatal journey, again some writer runs howling to his art.
Extract from Journey to Iceland by Wystan Hugh Auden.

In 1936, the writer and poet W. H. Auden visited Iceland, and I was delighted to discover that this great Yorkshire man had visited Blönduós. However, I wasn't quite so impressed after I read what he had to say about it. In Blönduós, he wrote that he was "served enormous hunks of meat that might have been carved with a chopper smeared with half cold gravy." I hoped the food had improved. Auden's family had Nordic ancestry, and he, too, had grown up reading the Norse Myths and Icelandic Sagas. Iceland was a holy place in his childhood, just as it was in mine. After reading Auden's letters, I didn't fully understand what he meant by "running to one's art". I decided I'd leave that to the textile artists.

On the seat beside me were my gloves, together with my camera and handbag, and some rough directions I had printed out just in case I had no Internet coverage. I needn't have worried. One thing that did work well in Iceland was the Internet. My directions were to follow Road 93 towards Egilsstaðir. I would stop there as the chap from the ferry company had mentioned the town was a reasonable size with a few shops and a garage. I calculated it might be as good a place as any to stop for the loo and maybe buy a few extra supplies (chocolate). When I travelled abroad, I always bought local chocolate bars, preferably with fun names. For years I had thought Toblerone was Icelandic (as my mother had given it to us as a gift after returning from Iceland). At school, an embarrassing discussion in geography class confirmed the Matterhorn on the packet wasn't in Iceland. Despite that fact, our mother had argued its false provenance and I gave her the benefit of the doubt in fear of being dumped on Surtsey. It was rarely seen outside airport shops back in the 60s and certainly not in the local corner shop cum post office on our Scottish island, so other than our cantankerous geography teacher, no one else argued.

The road ahead snaked up the mountain in a series of scary hairpin bends. Having changed my plans to arrive a month earlier, it now became apparent what a bloody good idea that had been. I would never have managed to drive my car up these mountain passes in ice or snow, probably even with studded winter tyres. I was not an aggressive winter driver, and my car was front-wheel drive only. On a dry day, as this day was, with the extra weight in the car, it was hard enough to maintain a steady rhythm of switchbacking along the ridges of the steep mountain pass. With snow and ice, I would have found it impossible. The wind didn't help. The strong headwind was slowing me down, and I had to accelerate hard just to keep moving. All I could think was keep going, don't stop or brake, no matter what, for fear of rolling backwards.

It was difficult to understand why the only ferry coming to Iceland was in such an inhospitable terrain. In summertime, it was surely manageable, but I couldn't imagine how, in wintertime, the roads could even remain open. I later found out that through the heavier parts of winter, they do indeed become impassable. I had been driving for twenty minutes and already started to worry how I'd get the car home again, or myself for that matter, come February or March. The essence of the word "travel" was travail, meaning hardship. I made a mental note to keep the shovel and torch very handy. Shit, the shovel. I remembered that I'd left it outside the back door and forgot to pack it.

No one could take away the fact that this was a stunningly beautiful panorama. Practicalities aside, it was jaw-droppingly impressive. The East-fjord landscape was far more dramatic than I had expected. No guidebook could prepare me for its beauty – not that I had read any for a good 20 years, to be honest. In my early days of travel, Baedeker often accompanied me. I loved the detailed maps and even the ferry and boat timetables, museum and historical buildings guide and information on local customs and manners. Over time though, with in-flight magazines and the Internet, I gave up buying travel guides. Not so many years ago, I flicked through one in a second-hand bookstore and realised how unarguably rude it was. With its derogatory comments, it was amazing to think the German captured the travel market guide for more than a century.

I had decided too against reading reviews about Iceland. I would make my own mind up about the places I visited and the people I met. I could honestly say nothing prepared me for what I saw before me. The immense mountains and cliff faces, lava fields and colourful flora were stunning. A rich bird life was evident too, and I imagined, in spring, it would be a frenzy of activity. It is, in fact, an ornithologist's dream to visit here in summertime, and I could fully appreciate why. On this occasion, I had no time to explore further, but I promised myself if the weather was fine for a few more weeks, I would return for a few days and camp.

Despite the tourism boom that Iceland has witnessed over the past couple of decades, parts of the northeast coast have remained a relatively well-kept secret. During my stay, I spoke to many people, and surprisingly, they said the same thing. On the whole, tourists ventured little outside of Reykjavik and the golden circle. Price and safety were factors and it was much easier to book up an arranged tour and be picked up and dropped off at your hotel or guesthouse. Nevertheless, in summertime, it was relatively easy to rent a car and enjoy the freedom of a road trip. Arriving by ferry as I had done took time so was not for everyone, though it did give you the perfect opportunity to explore this majestic coastline, and it felt so much more adventurous than simply arriving by air. Discovering the east coast truly was a journey of a lifetime where you would be rewarded with an insight into the real Iceland, its flora, heritage, and culture. You would certainly be fortified by the surrounding mountains and be captivated by the scenery waiting to be explored. Whether you enjoyed photography or history, you wouldn't be disappointed. Despite some villages being a little out of the way, there were many that were host to impressive collections of Icelandic history housed in tiny museums, for example, Natural History, a Wartime Museum, and a Maritime Museum. In fact, Iceland was saturated with small museums.

The road onwards to Egilsstaðir was decent and well surfaced, and as it twisted and turned, I imagined each bend must take me to the crest, but it went on, another view, another vista and then another. Finally, when I reached a point where there was clear blue sky beyond, I stopped the car and stepped out to stretch my back. I pinched my nose and swallowed hard to clear my blocked ears while feasting my eyes on the view before me. In the distance, the mountaintops were a patchwork of dirty snow. A vibrant autumn palette of lichens, heathers, and grass dressed the landscape, while the immediate foreground was a razor-sharp lava field, jet black and shining, looking as though it had spilled and solidified only yesterday. As I breathed in a deep lung full of fresh mountain air, I almost gagged with the stench of sulphur or rotten eggs. That smell would follow me around for months. The only sounds were the purring of the car engine and of running water. I looked around me and saw the waterfall nestled between two valleys. I sat back in the car, opened the glove compartment, and ate the emergency chocolate I had stashed there before leaving home.

I continued my journey, making a quick pit stop in Egilsstaðir as planned, and I really did make it quick for there was no temptation to hang around. Functional at best describes it. I was hungry and initially delighted to find a Subway open. I had never been in one before. The "sandwich artist" recommended a chicken teriyaki sub. However, I threw it in the bin on the way out, along with the coffee. I wonder what Auden would have written.

I was now officially on Route N1, Iceland's ring road that would take me all the way to Blönduós. As tempting as it was to venture off at times, I knew I needed to arrive in Blönduós during daylight and before the staff at the residency left for the evening. Perish the thought of sleeping in the car, or worse still, having to unpack my new tent (Bothy) on my first night here. Given the terrain, I once more got to wondering where exactly the interior started and finished. As I drove onwards, it was noticeable that this area was in the throes of change, no doubt for the hungry tourism market. This was understandable, given its proximity to the domestic airports of Akureyri and Egilsstaðir and the recent construction of spas and hotels. There was even talk of Akureyri opening to international charters soon. I was glad to be exploring before it became just another tick box destination on Instagram accounts.

I drove steadily onwards and through the Mývatn area, following the main road that would take me to Akureyri. Thankfully the wind had died down, and the sky was still clear. Driving through Reykjahlíð, on the north-eastern shore of lake Mývatn, I was enchanted. Itself a hub of guesthouses and hotels, a tourist information office, a petrol station and a grocery store, this small village was set against the stunning backdrop of craters, mountains, and the lake. I stopped the car at the first passing place I could find. Camera in hand, I switched off the engine, grabbed my jacket and hat and got out. A nearby information board gave me the basics. The enormous Krafla eruption back in 1727 set off a two-year period of volcanic activity. Today you could walk up to the Leirhnjúkur crater, around 10 km away on the northeast side of the village. It mentioned Dimmuborgir, and I remembered this from books we had been given on Iceland as children. Sometimes you must just stand still and drink in the views. I hadn't thought I would pass this and would return as soon as I could. It was not far from Reykjahlíð and should be on everyone's bucket list. I couldn't imagine a more impressive lava field, a natural art exhibition courtesy of the planet. Lava didn't get any more magnificent than this. The towers and columns were created when the lava dammed in a fiery lake, then the lake drained, leaving behind the oddities. Mythologically speaking, its home to the naughty Icelandic Yule Lads, and Christmas in Iceland wouldn't be the same without them pot-licking and window-peeping their way around the country. I remember my mother reading us the poetry book "Christmas is Coming" and relishing the verses on monsters with appetites for flesh and mischievous children, scaring me half to death.

The time was half past two, and I knew I had to get going as I had stopped for longer than planned. As I drove on, I passed a sign for Sigurgeir's Bird Museum and Café. I was desperate to pee and could murder a cup of decent coffee. I'd forgotten to pop to the loo at Subway and their coffee was undrinkable. A turf-topped roundhouse came into view, and I expected little more than a bird-watching hide. As I parked the car and walked the last few metres, the remainder of the building appeared: a modern, beautiful structure housing the exhibition, café, and shop. The lady running the business greeted me like an old friend and explained she was the sister of Sigurgeir Stefansson, and the museum was built as a memorial to him after he died beside the lake, aged 37. I shuddered as she told me their story. Sigurgeir was an avid collector of birds, and the museum housed his private collection of taxidermied birds and eggs. Every bird to visit Iceland was here, and its polished showcases rivalled that of any metropolitan collection I have visited. The small café served light food and I was fully expected to stay and eat.

Despite my protests that I was short on time, she recommended the rye bread baked in the ground with smoked arctic char and homemade pear cake. There was no point refusing and this stop was going to take more than the five minutes I had planned. After being fed, educated, and entertained with stories of the lake and its feathered inhabitants, I was guided to the last area, which housed a magnificent wooden rowing boat that serviced the lake for decades. I thanked her again, paid in the brand-new Icelandic Kronor I had bought before leaving home and promised to return before my journey north was over. She left me with a thought that piqued my interest.

"Next time you visit Ytri-Neslönd, Emma, I will show you where to find the rare Marimo balls."

I stopped and turned to her. "What balls?" I asked, with no idea what she was talking about.

"Next time," she said as she turned her back and quietly disappeared back inside.

As I drove away, I mentally noted this for the future. It was by chance, some weeks later, that I remembered our conversation and looked up lake-balls of Mývatn. Right enough, colonies of these unique balls grew here. They were balls of green algae that grew to the size of cabbages. The algae itself was not rare but forming itself into balls happened in only about three known places on earth: in Japan, Ukraine, and Mývatn in Iceland. Sadly, in the short time since I first heard of them, they have almost completely vanished from the area.

Driving from Sigurgeir's café to re-join the main road, I stamped on the brakes. I had won my second bet of the day with myself. Right there in front of me, sitting quietly with wide-eyed intent, was my first ever Gyrfalcon, the national bird of Iceland. I never expected to see one so soon and was delighted. I had bet myself that if I saw a Gyrfalcon during my first week in Iceland, I would buy an additional lens for my camera purely for photographing birds. I was delighted and knew that despite the challenges Iceland might throw at me, I had made the right choice in coming. I managed just a few photographs before the bird calmly took flight and settled again a few safe metres away. I couldn't wait to send Mats the pictures too, as he was an avid birder.

It was difficult when driving through Mývatn to concentrate on the road. The abundance of birds on the lakes and rivers even at this time of year was exceptional. I didn't recognise many species, though would buy a decent bird book before returning to the area. I'd have liked to detour and take the coastal route via Húsavík but made do with the direct Route 1 to Akureyri. I still had a good three hours driving ahead of me. The views were outstanding as I drove on and I muttered superlatives out loud to no one. Geothermal vents, mountains, rivers, sheep in all shades of chocolate, milk, dark and white, and horses with impeccably clipped manes and fringes. Taking the last bend high up the mountain pass, I finally saw the city of Akureyri below, across the water. As I pulled into a passing place for just a moment to take in the view, an enormous creature broke the surface of the water, and I found myself loudly shouting, "Whale!" with tears of joy springing to my eyes. I certainly hadn't expected that. I stepped out of the car and watched countless humpbacks breaching the water to breathe. Their flukes visible before going for their deep dives. The moment was surreal. They weren't shy in their playful displays, and I was astonished at how many there were. This road trip had already surpassed so many expectations and I shook my head in awe as I drove down the last hill and over the bridge into the small town. I thought of John and Mary and wished they, too, could have shared this sight.

Passing a petrol station, I decided to fill up the car with diesel, as I'd no idea what time the garage at Blönduós would open in the morning.

"Volvo," said the man in the garage when I went inside to pay.

"Correct," I said.

"You're from Sweden?" he asked.

"Yes," I replied, then, "no, actually I'm from Scotland, but I live in Sweden," I explained for clarification.

"What are you doing here?" he asked rather bluntly.

"I'm going to Blönduós," I replied, "I'm going to live there for some months."

"Why? It's shit," he replied.

That comment hit me like a slap in the face.

"Well, I'd like to spend time getting to know Iceland, and maybe learn to knit," I answered.

The young man stayed silent for a time, and as he was holding my bankcard, I couldn't leave. After a long pause, he eventually spoke.

"My sister will teach you. She's the best knitter in Blönduós."

"Oh." I hadn't expected that either.

"I will ring her now, but she talks a lot, so sit down."

"I don't have time," I said to his back.

I wandered around the shop, drank some free coffee, popped to the loo just in case and waited some more. He obviously hadn't exaggerated, and just as I had decided to go through the door marked private to find him, he returned.

"Here is her number. Her name is Ragna. Ring her when you arrive in Blönduós."

He handed me the note with her telephone number, together with my bankcard. All I could think was that I hoped it wasn't some sort of a ruse and he had copied my bankcard.

Politely, I thanked him anyway and got back in my car. I drove away, deciding he was probably honest, and if not, I knew where to find him and his sister. Maybe I would even ring her in a few days. I knew I would now miss the cut-off time for arrival, so rang the residency and explained where I was. They already knew. They had just spoken to Ragna; she had rung to say I was at the garage with her brother, and when I arrived, they must remind me to ring her. They would wait for me to arrive if I hurried up.

The road was decent quality though I was certain it could be difficult in winter to get around here. I started to wonder seriously if it had been a mistake to bring the car, but it was too late now. I couldn't sell it and buy a 4×4 either as customs would surely find out. Disappointingly, after Akureyri the N1 Ring Road didn't hug the coast, and the views were limited and less spectacular, still nice, but not a wow. It never takes long for new and exciting to become plain old familiar. As I drove, I became less impressed. The road narrowed and I concentrated on avoiding sheep wandering in the same direction. The landscape was spoilt by a large collection of plastic debris in the verges, likely cast out of car windows or blown from nearby farmhouses in windy weather. I studied the new road signs, including an alarmingly large snowflake on a blue background. Did it really snow that much? The signs had become more regular in this mountain area. As I approached turnoffs to gravel roads, the signs were ever-present. I eventually realised this must have been a sign for knitting patterns or garments. At least, I hoped so.

Seeing lights ahead, off to the right, I knew I must have been near Blönduós. A broken sign read "Hotel Blönduós". I was almost there. The N1 petrol station on my right confirmed this. I indicated right and turned off into what was described as the town centre with the buildings I had seen on Google Earth. On my right-hand side was the swimming pool with its large plastic slide, followed by a supermarket. The school looked bigger than expected. Attached to a lamp post with a brightly knitted cover was another huge snowflake and a sign underneath reading Kvennaskólann, which directed me left. I felt nervous and needed five minutes to compose myself. Despite my lateness, I decide to drive once more around the town, checking my pockets, my handbag, my phone. I tried to hold back the tears but could feel them pricking at my eyes. What happened to the beautiful towns I passed along the way? For the first time, I seriously didn't think I could go through with this and started to panic. The ferry would be leaving for Denmark tomorrow morning, and I could make it. Could I really stay in this godforsaken place? Ragna's brother in Akureyri wasn't wrong. It looked shit. As I drove to what I hoped was the harbour and beach area, I found only the slaughterhouse and rusting containers. I made a three-point turn and headed back out the way I came, crossing the bridge to the other side of the river. I passed the police station, hospital, and post office and came to the closed-up Hotel Blönduós, which I was sure was the hotel my sister had booked in advance for when she planned to visit me. More rusting containers and debris lay strewn about the car park and outer areas. *It's where containers come to die*, was all I could think. Truthfully, it was fucking horrible. A real shit hole and I had foolishly paid to stay here for six months.

The car clock read 17:40 and I felt sick. I decided I must drive to the residency as they were waiting for me. I'd tell them my plans had changed, let them keep the money I had paid for the room and then drive back to Seyðisfjörður and catch the next ferry home. The problem was I had sold my house. Everything was in storage. I had nowhere else to go, at least not in a hurry. Despite the hurdles, I'd work something out on the way back to Sweden, I told myself. I wiped my tears and felt calmer with my new plan. I drove again over the bridge and found the artist residency, a large white building at the end of the road, next door to the abattoir. I parked beside the concrete block of garages with their red corrugated iron roofing and made my way to the front door of the main building. It all looked and smelt rather underwhelming, but that was fine; I wasn't staying.

I was greeted by a large smile and even larger warm hand. "Welcome to Blönduós, Emma. Now hurry up and follow me as I must leave for a goose party, and you are very late." She took my hand in hers, and I found it impossible to get a word in. This was Jóhanna, the director of the residency and one of my extremely patient knitting teachers, or so I had hoped.

Two faces appeared along the corridor, and I felt as though I was the newest inmate of a prison. I had to speak up soon. Jóhanna insisted on showing me the view from my room. I saw the garages I had parked beside together with several grim industrial units and a dark mountain beyond that, nothing more. The room had two worn-out single beds, a ratty dining chair placed between them with a lamp that had seen several decades. There was no writing desk. I asked what had happened to the room I had been promised, the one with a lovely view and large desk, which I had paid a generous supplement for.

"This is not available until the end of next month, so you will have to make do." Jóhanna shrugged.

Across the hall, a smile appeared around a door and a brief welcome. A young lady, red-eyed from either tiredness or tears, greeted me with an American accent. She explained she, too, had just arrived a few hours ago by bus from Reykjavik. Having introduced myself, I turned back to speak to Jóhanna, but she was gone. I hurried back down the stairs and out of the door but saw only red taillights disappearing round the corner.

The American lady was waiting beside my door as I returned up the stairs. She asked if I wanted to go out to find something to eat.

"I'm not staying," I replied.

"Oh! Lucky you, I haven't a choice," she whispered. I felt instantly sorry for her.

Perhaps we could go out to eat and then I could get back on the road to Seyðisfjörður. A little company would be nice, I decided. We headed out by car and found the only option was the N1 petrol station and diner, which would be closing in an hour. The grill served hot dogs, pizzas, burgers, and sandwiches. The "daily specials" offered meat soup, and I chose that; however, "special" was an over-statement. I was served a gooey mush of packet green mix, accompanied by a slice of processed white bread. Apparently, the homemade lamb soup had sold out. My American friend ate her hot dog quickly and moved on to coffee. Like an unhappy child, I played with my food, then pushed it aside and sulked. My heart was breaking, and I had no one to turn to. I chastised myself for my unrealistic expectations of how this trip would be and wondered, not for the first time, why did I hang on to romantic notions from my mother's past or wish to be so different from other people?

Suddenly, raised voices and waving caught our attention. A group of American ladies in the far corner of the diner were asking, "Are you staying at the residency or just on holiday?" I could see my new friend relax immediately and knew she'd be fine. I, on the other hand, knew that I wouldn't be. We joined them and they bombarded us with questions about our art, then filled us both in on details of the latest gossip, who they liked and disliked, all the while embroidering and knitting. When they learned I was a writer and not a textile artist, they lost interest in me. My new friend turned out to be a rather well-known New York-based artist with connections, and they understandably adored her. She was the sweetest girl I had met at the residency, and her craft was rather brilliant and inspiring.

Artist residencies weren't for everyone, but they did serve a purpose. Kvennaskólann in Blönduós opened some years ago to house predominantly textile artists from around the world. All living and working expenses were found by the artists themselves, including their travel and transport, materials, food and living needs. Iceland wasn't a cheap place to live, but it surely was a place of inspiration that no amount of money in the wrong location could provide you with. It offered the opportunity for both personal and professional exchanges between artists, and the whole objective was to support artistic research while showcasing the area and its rich cultural heritage in the art of textiles and farming. The whole operation was housed in the historic lady's college or Kvennaskólann, which was the large white building I had booked a room in. In daylight, they assured me, the building offered an imposing viewpoint on the river Blanda with a partial view out over the sound. Despite its limited perspective, it was in an ever-changing tapestry of a landscape. The near view was all you really saw, but they all agreed that for an artist's eye, the view and the light quality were never the same from one hour to the next.

As the garage was closing around us, we made a move. On the way out, several ladies confided they, too, had experienced panic on arrival. They planned to stay one month. None had the luxury of a car, and all had made the journey by bus from Reykjavik. In fairness, they did a great job of cheering us both up and recommended that I gave it a go. One or two of them had been there some weeks already and confirmed they spent most of their time indoors but liked to come to the garage for dinner occasionally. Buoyed on by their helpfulness, and as I'd already paid, I decided to stay a few days and give the place a chance. The road to the ferry was no problem just now, and there was no snow expected any time soon. Back at the residency, I was left alone to unpack the car and cram my boxes, spare wheels, clothes, and equipment into the corner of the tiny room to the point they almost reached the ceiling. The girls had assured me that in one month I was to get the bigger room. *There was no chance I'd be staying that long*, I thought, but said nothing.

I grabbed my wash bag and headed off down the corridor to explore the bathroom, passing by the kitchen on the way, which turned out to be basic and rather dirty. No, in fact, it was filthy. I wondered if things could get any worse. Several more people were wandering around, and I noticed that so far, it was only women. They were mainly very young and from the US and Canada. One European accent caught my attention. I reminded myself that what didn't kill me made me stronger. Morning would come, and with it, choices. I slept surprisingly well in the uncomfortable single bed.

GRAPEVINE

Up and dressed early, I went downstairs to the office to learn their director would be absent for a few days. The weather looked fine, so I decided to take an early morning drive to see more of the area. Surely it must look better in daylight. I joined the N1 where I had left off the evening before and, hoping to put some distance between myself and Blönduós, drove for over three hours and found myself unintentionally on the outskirts of Reykjavik. An hour or so from Blönduós, the terrain had changed, the houses became prettier again and the sun shone. My heart lifted and so did my mood. I wondered if I could handle living in the old house with the artists if I spent as much time as possible outside and exploring the rest of the country. In fact, I might even enjoy myself. My initial impression of Blönduós and the residency had been a huge disappointment, but my overall plan didn't necessarily have to change.

I knew I couldn't stay in Reykjavik and didn't want to. As I mulled through my options, I noticed the clouds were darkening overhead and the wind was picking up. I resisted driving into the city centre and instead filled the car with diesel and again headed North for Blönduós. One of my sisters planned to visit me in six weeks' time despite my being against early visits as I wanted this trip to be about me, but I now felt this could be my marker. Stay until then, if only to prove I was not a complete failure and if I still felt unhappy, I could pack and leave after her trip. Mind you, I would have to find better accommodation for her than Hotel Blönduós. She was a hotel manager herself and would sleep in a ditch rather than in a shabby hotel. Mats also talked of popping over, but I knew he would expect to camp outdoors. Mats, of all people, I didn't want to let down, given his help and support; he'd be disappointed if I threw in the towel without even trying.

Driving north, with time to reflect, I felt lonely. Being in a strange place with no friends or family around me suddenly made me feel vulnerable. Maybe I really wasn't a self-confident, no-keys kind of person after all. I noticed the wind, too, was getting stronger, which didn't help my confidence. What would I do if something major happened like a volcanic eruption? The toxic gases couldn't be good for asthmatics and the hospitals would surely prioritise locals. How would I escape? I wished I'd taken more notice of the news in recent years. I knew Iceland had suffered a few major volcanic eruptions which disrupted air travel, but that was about it. I wasn't enjoying this experience at all, and to comfort myself, reached for my handbag and went ice-cold. It wasn't there. I glanced at the passenger seat, empty. Phone, purse, spare car key, passport, driving license and insurance documents, money, credit cards, inhalers for my asthma. Oh God, please no, don't let this be happening. "Don't panic," I tried saying to myself. "You've a full tank of diesel so can drive back to Blönduós and sort everything out." Had I taken the bag out of the car at the petrol station? I didn't think so. Usually, I just took my purse, but I couldn't be certain. I had left the car unattended for a few minutes in Reykjavik. It must have been stolen. Fuck!

I braked, pulled over into the rough verge and, seeing headlights behind me, turned on my hazard lights. The car behind me slowed and stopped. I opened my car door, well I tried to open it, but the wind was so fierce I could barely force it open. A figure appeared and took control of my door. I stepped out and thanked the man for stopping. I wondered if he was dangerous, then instantly decided most strangers were kind. He had, after all, pulled up to help, hadn't he? I would be meeting many strangers on this trip so I must learn to trust people, though after my break-in, I found it hard since the police said it was likely someone who had watched and followed me, or worse still, knew me. All I really wanted was to be alone so I could search under the car seats properly, despite knowing there was no real chance my bag had slipped underneath.

"Do you need help?" asked the man with long hair and an even longer beard.

"No, I don't need help but thank you for stopping," I replied.

"I have been following you," he said in broken English and smiled.

I didn't reply. I now wanted him to get lost.

"It was fun following you, you drive fast. You stopped your car when you saw me," he said.

"Look, sorry but I don't need help, and I didn't stop because I wanted to talk to you," I explained trying to appear in control and unafraid.

"You are nice girl, from Germany?"

Oh shit, here we go, I thought.

"No, Sweden," I now snapped back.

"Yes, good Swedish car," he replied.

I turned to ignore him. "Come to my car. I have something you like," he said as he firmly held my elbow.

The wind blew hard, and I barely heard his words over the heartbeat in my ears. I tried to move away, but he was determined. I looked up, and there was at least one more person in his car. My knees almost buckled when he looked straight at me and smiled.

"Come on, meet my friend, he likes to talk English."

More forcefully, I now turned to him. "Look, I'm not interested. I only stopped my car as I thought I'd lost something but now I need to get home. Someone is waiting for me, and I am very late." A lame attempt, I admit, to act as though I was unafraid and not alone.

He mentioned something about money, but over the wind and the music coming from his car, I couldn't hear. He still held my arm with one hand, and as he opened his car door with the other, I glanced inside to see his companion smiling and holding my handbag. I almost fainted with relief at not being attacked and more so at having been reunited with my bag. Apparently, they had followed me for several kilometres out of Reykjavik in order to return my bag that I had left behind in the garage. I remember now I had popped to the bathroom, and the floor was wet and dirty, so I hung it on the hanger at the back of the door. I had already paid for the fuel. It was an old-fashioned garage where the attendant had filled my car up while I went inside to pay the cashier. This was the first time I had ever lost my handbag.

These two young men had volunteered to go find the woman in a Volvo with a Swedish number plate. I had mentioned in the garage I was driving to Blönduós, and there was only one main road to take. This act of random kindness surprised me, and furthermore, despite me trying several times to offer them a reward, they would not hear of it.

"Absolutely not, but if you are going to Blönduós could you say hi to my father's cousin, he washes wool there," said the driver. I promised I would. They gave me a warning about driving too fast as, after all, I was only a tourist and when I did attempt to defend myself, explaining that I had driven for several years in Sweden during winter, the two men replied, "Ha!" in unison.

"Keep looking at the Vegagerðin app," they said. "It's your best friend."

With that warning and a wave, they drove off. As I struggled against the wind to open my car door, I wondered what they meant. I wrote down the name of the app they had recommended and would look later. I couldn't believe I had mistrusted these guys instantly when they had been so kind. Hopefully, my rudeness was lost in translation.

I hadn't planned to drive so far, and by now it was late. The drive back was around 240 km from Reykjavik to Blönduós and would take me a good three hours. I'm not sure what I had expected from the road conditions in Iceland, especially the main ring road, the N1 as it was known, but something better than what I'd so far experienced.

A side wind blew ferociously, and I concentrated hard on driving, my knuckles white. I had driven through Hvalfjörðurtunneln for the second time that day, 5762 metres long and its deepest point 165 metres under sea level: Hvalfjörður was the closest major fjord to Reykjavik and was, during the Second World War, used by both the British and American navies. The Royal Navy assembled the Murmansk Convoys here. Today one of the old naval piers was used by Hvalur, the only Icelandic whaling company in existence today, for the processing of fin whales for domestic consumption together with the Japanese market. Onwards, over the Borgarfjarðarbrú, the second longest bridge in Iceland, crossing over the muddy waters of Borgarfjörður fjord and through the town of Borgarnes. I didn't stop. Many early settlers in Iceland called Borgarnes home. I simply wanted to reach Blönduós and the safety of the residency with the girls and their positive humour. Maybe I had little in common with them, but being part of a community suddenly felt more appealing than being alone. I was glad to have the independence my car gave me and was even happier to have my handbag! Sitting in my car felt like familiar territory, and at least it was safe. As the kilometres passed, the wind increased. This could be an adventure of a lifetime, I told myself, but I knew I had to be more careful. This was a journey of choice: there was an entire island waiting to be explored, and I knew I must take each day as it came, give it a chance. Forgetting my bag was careless and unnecessary. Coming to Iceland wasn't totally impulsive; I had thought things through and planned as much as I felt I could in the final weeks before coming here. Now I was here, I couldn't be immature and just give up. Having given myself a stiff talking to, I drove onwards feeling more positive. I had had a shock mislaying my bag, but it wouldn't happen again. I'd make sure of that.

I passed new road signs with place names that kept me entertained. I started to feel back in control and noticed much more than I had earlier on in the day, making mental notes of the places that interested me. Quite why I had been so shocked at my first impression of Blönduós, I had no idea. Perhaps it was just tiredness. Perhaps my expectations were too great. Either way, I promised myself I wouldn't let it happen again. So what if I needed to clean the accommodation regularly, or that it was basic and threadbare? Who really cared if the plastic shower cubicle had black mould in the corners, or if the drain was blocked with years of neglect? If I had to stand ankle-deep in warm, soapy scum, so be it! I reminded myself I had been forward-thinking enough to pack my own my own crockery and utensils, and a few luxury items that having the car had afforded me. What about the artists who had to stick to luggage allowances with working materials taking up most of their weight limits? Of course, a few probably didn't even notice the living conditions, and that was their prerogative. If I could arrange a desk and chair to work at, it would be fine. I turned up the car radio to drown out the wind and drove onwards, south to north on the N1 ring road. The signs advising me of the approaching summit height had red warning lights showing the wind speed and direction. I wasn't sure if 28 metres per second was anything to worry about but decided if it was in red, it must have been strong. I hadn't noticed any warning lights earlier in the day. The car rocked against the wind. I pulled over to take out my camera, and while I was safely pulled into a passing place, took the opportunity and searched "wind speed" on my phone. Apparently, 20 metres per second was a gale force. That gave me something new to worry about. Just how windy could it get here?! I could tell you; it could get a whole lot worse. 28 metres per second by the end of my trip felt like a gentle breeze. I placed my phone back in my breast pocket of my fleece jacket, a precautionary measure Mats told me to always observe. He persuaded me that if I was involved in a crash and broke my arm, it would be easier to access the telephone for help. As I drove onwards, I silently prayed that winter proper would stay away for a while at least, allowing me to get about and see a little more before having to decide how long I would stay in Blönduós. As I crawled. up the steep mountain pass, the wind increased further and there was sudden fog, which was so thick I couldn't see the summit, though knew I must have been almost there as there was nothing to shelter the car from the buffeting. I held the steering wheel more tightly and dared not blink for fear of missing the edge of the road. The small lake to my right was having difficulty holding on to its contents as the wind blew the water into an airborne wave. I gasped. I had never witnessed such force before. Thankfully after around 40 minutes of the worst driving conditions of my life, I came down from the high ground and pulled into a welcome sight, the Staðarskáli restaurant beside the N1 petrol station, just past Staður. I stepped out of the car with heavy legs and aching arms, feeling as though I'd had a workout. I took several gasps of fresh air and bolted inside for coffee. Checking the map on the wall, I was still around one hour from Blönduós. It was sheltered here, and I found it hard to think that minutes before to the south, that schizophrenic weather was in full swing.

Parking outside the concrete block of garages in Blönduós with the red corrugated roof, I felt relief at having made it back in one piece. The three-hour journey from Reykjavik had taken a full five hours and not without drama. I sighed deeply, locked the car, and went inside. "Welcome home", I muttered to myself. Heading up the stairs, a jolly and rather large Icelandic woman chased after me.

"Do you have your purse?"

"Yes," I replied, "but how did you know?"

"My husband's colleague's friends met you. They rang to make sure you hadn't lost it again on the way home. Haa!"

I turned and hugged this stranger. I suddenly didn't feel alone at all.

LEARNING TO KNIT

Pulling out the piece of paper I had popped into my pocket some days earlier, I read the telephone number Ragna's brother in the garage in Akureyri had given me. I had promised to ring his sister and enquire about knitting lessons. So much had happened in my first few days I had totally forgotten to contact her. No time like the present, I decided and dialled her number. We planned a meeting for the following day.

Walking the streets to my first knitting class, despite the drizzly weather, I was excited. I had every size of needle I thought I could possibly need, together with a variety of wool and patterns in three languages. Having bought some wool and knitting needles in Tórshavn en route to Iceland, I had now added more to the collection from the local supermarket. I bought a large quantity of the traditional Icelandic *Lopi* wool on the recommendation of one of the artists and felt well prepared for my first lesson. I hadn't considered how the wool felt against my skin and quickly realised I didn't like the scratchy feel of it. Also, I realised it made me sneeze and itch. I rushed to the shop at the last moment and restocked with Baby Merino wool and a new pattern for a hat. The sweater, cardigans, and socks I had dreamed of knitting would have to wait until I found more wool.

The welcome I received was like coming home, as Ragna opened her door and greeted me with a wide smile. On her shoulder was Kopi, her pet parrot who rarely left her side. Their home was traditionally Icelandic, similar in ways to the home I grew up in. Nothing was thrown away and everything was useful or cherished and had a story. Furniture was polished, cushion covers were knitted, elegant lace curtains were handmade, tablecloths were crocheted, and fruit bowls were also hand-worked and sugar-coated. The kitchen was no less industrious. An array of jams, including crowberry and arctic bramble, were laid out on the table for tasting together with locally made cheeses. The generosity extended here was bountiful, and I was urged to feel at home, and so I did. Her English was rusty, but her generosity and kindness were certainly not.

If only I could hit pause and rewind as I watched Ragna's fingers move like a machine. She tried her best to knit more slowly to allow me to watch her technique though I was too busy marvelling at her skill to be able to figure out how to copy her myself. I realised that learning to knit was going to take time. Ragna assured me it was easy once you got the hang of it, but I wasn't so sure: it was only after three hours that I had managed to cast on and start a row of actual knitting. However, her circular needles were also new to me, and suddenly, I broke one. I would have to start again tomorrow when I had bought a replacement. Having talked and laughed more than we knitted, I promised I would try harder next time. Time to make an exit and head back before the dark night set in, I sneezed all the way home. I hated to tell Ragna that I suspected I was also allergic to birds and, by evening, was taking my inhaler to ease my wheezing chest. Two more lessons with Ragna, and it was confirmed: my allergy was down to the parrot, and I couldn't continue. The Lopi wool was merely irritating. The bird was rather more worrying. I'd have to rethink my knitting lessons.

Since my arrival, I hadn't caught up again with the residency director. Three weeks had gone by quickly when I heard a singsong voice bidding me good day. We chatted, and she was concerned how my knitting was going.

"Good day, Emma. How is the knitting coming along?"
"Jóhanna, lovely to meet you again. Slowly," I replied.
"That won't do. Come to my office at five o'clock."

I hoped I could also ask about the larger room I had paid for. Over cake and tea, we knitted and talked, and I told her of the notebooks I inherited and why I had decided to come to Iceland. In fact, just days before, I had re-read some pages from my mother's time spent in France with a family who were involved in working on the backcloth for Coventry Cathedral in around 1962. I didn't know a great deal about it other than what was written in her journals, though Jóhanna turned out to be incredibly well-informed. She went on to tell me her life story from being a past pupil of the ladies' school here, her time living in Denmark, her years of teaching embroidery, and of her vision to make the world's largest single piece tapestry.

As I was interested, she forwarded me some information. "Vatnsdœla saga is the family saga, written around 1270, of the people of Hof and takes place during the 9th to 11th century. It is a story of fate, love, and honour as well as struggle against dangerous enemies. The main setting stretches from Norway across to Vatnsdalur (water valley) in Iceland, around 20 minutes' drive south of Blönduós." Jóhanna's vision was to make a tapestry depicting the saga in the style of the Bayeux tapestry, which was sewn in the 11th century. She embarked on this enormous project some years ago, and it was now well underway. Under the leadership of Icelandic artist Kristin Ragna Gunnarsdóttir, along with second year students from the graphic department of the Icelandic University of Arts, the tapestry was drawn up in its various stages, and was now securely housed inside the residency where, under the watchful eye of Jóhanna, it was worked on by people from around the world. Visitors could work on this piece of history, paying by the hour to do so. The money went towards funding the project. When the tapestry was complete, all 46 metres of it would be on permanent exhibition in a purpose-built facility beside the old church at the foot of lake Húnavatn. Húna meaning polar bears young, and Vatn the water at the mouth of the Vatnsdalur valley. Lying across the lake to the north was Jóhanna's farm where she lived and ran a sheep farm. I was sure for generations to come, there would be a close eye and stories added to the sagas, keeping tradition alive.

"Did you know that Eric Clapton fishes here?" she asked me. Apparently, he had not yet met her. He should. She was an inspiration on many levels, and she could help make the valley he was spending time in come alive with her wonderful tales of every hill and stream. One thing was for sure, my knitting might be taking a little time to come together, but I was going to sew some tapestry. She assured me that anyone could learn, and I booked some hours immediately. I noticed on the tapestry a rather beautiful mother polar bear with two cubs, so far unembroidered. Polar bears were my favourite animal, and while the last dream I had has so far kept me from opening my tent, I certainly was excited about sewing one into the history books.

"You will start to sew the tapestry tomorrow," Jóhanna told me. "Now we need more cake, and don't forget to knit," she commanded me.

AROMATHERAPY

The joy of stepping outdoors into the early morning arctic sea air was fortification for your lungs and soul. That was, if the wind was in a favourable direction. If not, it could be an assault on your nostrils. Being uninitiated to living so close to a slaughterhouse, or Kill House as it was known locally, I realised, all too late, when I had inhaled a lungful of air that it was filled with a putrid stench. Blood, meat, bone, with a tinge of crematorium smoke. I struggled to exhale with closed nostrils and mouth and buried my face deep within my scarf to stop myself from projectile-vomiting over the nearest lamppost "stocking". Blönduós's lampposts were beautifully decorated with yarn bombing left over from the recent knitting festival and I didn't wish to vomit on them. The weather was plus 8°, and large bluebottle flies were in every window. Now I understood why they thrived here.

A few jaded starts saw me venturing out in the morning air. Tentatively sniffing, checking the wind direction, and depending on the level of funky air, decided in which direction I would walk. I asked the locals what they thought about the smell.

"Unpleasant occasionally, but tolerable," said one.

"What smell?" said most.

Getting to know this small town didn't take long. The walks over the mountains on the outskirts of town brought you in circles, so it was difficult to get lost. Such walks were a welcome respite in the afternoons when the residency became busy and elbowing your way into the kitchen wasn't pleasant. As I stopped at the local café, it was awash with foreign accents. I was surprised to see so many tourists, young people especially, venturing this far north so late in the season. Most were simply driving the Ring Road with a few stops at the better-known and prettier towns along the way. It was easy to strike up conversations, and everyone had stories, usually about the weather. Mixed with the tourists were another group of accents, a transient workforce here for the season. New Zealanders, Polish, and Filipino supplied the extra manpower required within the Kill House and fish processing plants, returning yearly like migrating birds. Work was available and the pay was good, they told me. The local off-license (Vínbúðin) certainly didn't complain as they benefited from the enthusiastic new customer base. In fairness, when we did hold an exhibition of our work in late October, the manager was the first one to visit us in support, in fact, he was the only one for the first hour and a half of the two-hour long exhibition.

A frequent companion on my walks were the rusting, sinister-looking trucks collecting their next victims for the Kill House. It was as though wherever I went that factory followed me. The transport truck, with ventilated siding, tumbled through the country farm roads and onwards through the main street to its final destination, the Kill House. The holes on either side allowed just enough space for me to see the sad eyes and pink noses taking their final breaths of fresh air and look at the world. The animals had a good quality of life in Iceland, roaming free and enjoying an organic diet for the most part. I tried to make myself feel less troubled by that but found as the days went on, I was buying far less meat here, despite it tasting so good.

Heading back from a long walk, my phone rang.

"Are you at the school?" a woman's voice asked.

"Yes, almost," I replied. Only locals call the residency "The School".

"I'm outside."

"Okay, I'm on my way."

It was Ragna. Since the episode with Kopi the parrot, we met occasionally outside the school to walk, talk, or take tea. She had come to invite us artists in residence to visit the elderly community who met twice a week to make handicrafts, chat, play cards and enjoy one another's company. This was such a sociable and welcome gesture, and I, for one, wouldn't refuse. I promised to pass the word around and get as many others as interested as possible. I often heard that residents didn't interact in the community and thought this would break the ice. The following Tuesday came, and five of us made our way to the Community Centre beside the hospital. Tea was served with plenty of home baking, including the Icelandic favourites Kleinur and Ástarpungar (Love Balls) which, given the name I was suspicious of, *pung*, meaning scrotum or sack. Despite the lack of a common language, the locals were smiling and welcoming, showing off their skills acquired over a lifetime. Everyone was equal, and everyone was welcome here with the sole purpose of enjoying the company of others while working their hands and minds. One lady who was in a reasonably advanced stage of a form of Alzheimer's disease was gently cared for by the rest of the group, and the patience shown to her as the carers stitched notes onto her work as a reminder of what she should do when she went home made me smile. On these days, Ragna put her knitting needles down and played cards with the non-knitters. I was thankful a handful of the girls from the residency followed me, though after a short time, they left, and I was alone with the group.

Barely three weeks had passed since I had arrived in Blönduós, and I now knew the town intimately. I knew each dog walker's routine, even the dogs' routines, which mutt liked which rock to take its first pee of the morning. I, too, was a creature of habit, so I sympathised with them. I knew the houses that cleaned their windows and who didn't, who picked up the rubbish from their garden and who left their muddy boots outside. The houses themselves were weather-beaten. The Icelandic climate was responsible for a multitude of building defects, which gave the whole town, and in fact, entire communities, the look of dilapidation. The concrete cracked under the strain of salt, wind, and cold and was left like a broken vase crudely joined with putty. Many houses were built for functionality not beauty, and I wondered why go to the trouble of repairing them at all. I made a mental note never to engage the services of an Icelandic architect. In their defence, I supposed by the time they considered the weather, earthquakes, and material costs, there wasn't always a budget left for aesthetics. I started to forgive Blönduós for its shabby exterior. The town was simply wind-bashed and weather-beaten, and I was starting to show the effects too. My skin was dry, my lips were chapped, and the weather was turning colder by the day, much colder.

SKAGASTRÖND

By late September, I was settling into a routine. I took a long walk in the early morning, then ate breakfast at the residency. I sat down to work for a couple of hours, then once it was daylight, I would explore the area with my camera, sometimes on foot and sometimes with the car. Around five o'clock, I would head back to the residency to join our newly formed knitting circle (prjonakaffi, aka knitting coffee) as it was known locally. After dinner most evenings, I dressed warmly, left the building again with my camera and car keys and drove out of the small town with no fixed plan other than to hunt for "the lights". The artists I shared accommodation with were late to bed and late to rise, and for this, I was thankful.

One evening a couple of the ladies asked if I would drive them to Skagaströnd, a small village around 20 minutes' drive from Blönduós, as they wanted to visit another art residency. I was surprised there was more than one, and although I had no interest in visiting it, I was happy to be their taxi for the evening. One artist residency was more than enough for me.

It was a perfect reason to go to Skagaströnd, as I was already a little curious about this village. It offered something out of the ordinary. A witch museum wasn't everyone's cup of tea, and it hadn't been high on my list either until I heard about the palm and rune stone readings. Although I liked to think of myself as not easy to bullshit, I was curious like the next person, and the truth was I did want to believe in the power of foresight, and so naturally, I couldn't wait to get my cold hands read. I immediately agreed to the girls' request and rang to make an appointment with the Museum of Prophecies later that same evening. We drove through the small fishing village, which had a slightly quainter look than Blönduós, which was not difficult, and I dropped the girls off at the NES Artist Residency to enjoy their potluck evening. I then had a short while to wait till my own appointment. Skagaströnd was on the very peninsula where a polar bear came ashore some years ago. It died of disease before it could be shot by an organised team of polar bear assassins, or so the story went. Its taxidermied cadaver was the one on display at the sea ice museum housed within Hillebrandt's House in Blönduós. The bear had been named Birna, after the young girl who'd first spotted the bear across the fields and mistook it for some plastic – until her dog barked in distress, and truth be told, she feared more for the dog's life than her own. Luckily a farmhand from Greenland working on the farm understood immediately what it was and firstly led the dog away to settle the young girl, and then guided her to safety. This, by chance, was one of the first local stories I'd heard on arrival at Blönduós. Little did I know that one of the ladies I would meet later that evening at the spooky museum was related to that little girl, who was now a grown woman with her own family. It was apparently not the first nor the last polar bear to swim up to these shores, although there seemed a difference in local opinion as to how many and when. One story I heard a little later involved some French anglers who had raised the alarm when they thought they had seen a polar bear while out fishing for trout. Following an extensive search from land and air, the search was called off, and the culprit deemed to have been a large sheep. Given that there were more sheep than people in Iceland and that the men ran four kilometres to their car over rough terrain, they certainly didn't think it was a sheep. Thoughts rushed back to my Hilleberg tent, still unused, and I doubted that would change any time soon.

I sat on the vintage velour sofa and waited my turn. A couple of middle-aged ladies beside me sat in deep conversation while knitting. I, too, had taken my knitting, which resembled an over-handled piece of dishcloth. After seeing these women drum their way through row after row of elegant stitches and intricate patterns, while managing to hold a conversation, I chose not to reach for mine. I tried to make small talk instead.

It turned out that one of these women was also a palm reader, rune reader, card reader, and coffee reader. She didn't feel her English language was good enough to give readings to foreigners, although I personally thought she spoke it very well. As we waited, her simply passing the time and chatting, me waiting for my scheduled appointment with a gifted reader, she placed her knitting slowly down and asked me to close my hands. She then took my right hand in hers. She said she felt the desire to "read me, just a simple palm reading to pass the time before my appointment." She didn't want payment. As she studied my hand, she explained the length of the fingers and the dominant lines, what they meant and how they could affect our lives. Our characteristics, tendencies, and humour could all be unravelled by the simple act of reading the palm, if you knew how. My stubborn streak and impatience didn't stay a secret for long, to the sniggers of the other witnesses in the small room. She went on to say I would continue to lead an interesting life though not always have a clear or easy path. There was unexpected great heartache around the corner, but also new beginnings ahead. She recommended my perfect job was a teacher and that I should learn to listen more. I took what she said with a pinch of salt and thanked her. Anyone could guess a woman of my age in the far north alone in winter might be going through some sort of changes in her life. The "professional reader" whom I'd booked called out that she was ready, and I walked through the saloon-style doors into the private reading rooms, off from the permanent exhibition space. There was a menu to choose from, so I chose runes, with a card reading if we had time, but left the tarot for another day. I also got to ask a yes or no question of the Vala. I'd never heard of this before. Vala was a bone from a sheep's knee used by Icelanders to answer questions to which they absolutely needed to know the answer. You held the Vala in your hand, said the rhyme and then dropped the bone on the floor.

"Answer me, dear Vala, the questions I will ask; I will bestow on you gold for your delight and silver for your desires if you tell the truth. But throw you on the fire and drown you in the chamber pot if you tell me lies."

Now, if the Vala dropped convex side up, the answer was "yes" and if the convex side was down, the answer was "no". They say that if it landed on its side, either there was not a yes/no answer, or it didn't wish to give you the answer. I asked if an impromptu best of three worked, but my reader didn't appreciate the humour.

For the equivalent of around thirty quid, she certainly did a good job of boosting my ego and confidence, and I was left feeling there were some great opportunities ahead if I just looked for them. As I left, she ran after me to ask: had I been christened? If not, had I drunk much horse's milk as a child? There was a treasure chest, apparently waiting for the right fair maiden to come along who shall be guided by two Ravens. They held the key to the chest, but I informed her sadly, I didn't qualify.

My scheduled 20 minutes had run into 40, and with barely a rune left unturned, I plodded out of the door, wondering what else the local knitting crowd had heard from behind the partition.

Once outside, the first lady who had given me a quick reading suddenly walked towards me.

"Emma," she said, "if you didn't want to hear the truth then you should not have held out your hand."

"No problem, it was fun," I light-heartedly replied.

"Remember you are strong," she said as she turned to walk away.

I thought her a bit intense but cheerfully thanked her again and said goodbye. Maybe she was hoping to be paid after all. She stared back at me over her shoulder without a smile. It was a little creepy, like she knew something I didn't! The second reader had been much jollier.

The girls were in high spirits as they climbed into the car, looking as though they'd enjoyed more than a few beers. I couldn't help being a little envious of how they had clicked and become a tight group. I was on the outside looking in. They were eager to hear about my evening, so I whitewashed a version for their benefit. We laughed the short journey back to Blönduós.

I declined their invitation to drink wine in the kitchen as I desperately wished to sit and write before I forgot details of the evening's events. I had also decided that I was spending too much time in Blönduós and, in the next day or so, would start exploring further afield and therefore had packing to do. It had been good to settle in a little and get some bearings, but now I was ready to see more of Iceland. I set my camera batteries and telephone on charge, pinned my map of Iceland I had bought the day before to the bedroom wall and decided to light a candle. When we drove home from Skagaströnd, there had been some cloud in the sky, but I had checked the aurora forecast, a daily habit by now, and there was a small chance later of seeing "the lights". The candle flickered, and as I sat down to write, the phone rang. I wasn't expecting a call, but occasionally, family or friends rang and asked if I was available to FaceTime or Skype. It was a UK number though not a number I had saved on my new Icelandic SIM card. I answered and was thrilled to hear it was Shauna, Mats' sister, whom I hadn't spoken with in ages. She worked in aviation, based in Dubai, and with a full-time job and kids, we caught up less and less. She sounded as though she had a cold. As the call went on, I realised this was no social call. I stopped talking and listened. My veins turned ice-cold, and I barely heard the words through her tears. Mats had been killed in action. She mentioned something about an IED in Helmand Province, Afghanistan. She had more details, but they would have to wait. I couldn't absorb more than I had. We agreed to talk the following day, and I concentrated on breathing through the feeling of drowning.

MANUS I MANO

Few hearts like his, with virtue warm'd, Few heads with knowledge so inform'd; If there's another world, he lives in bliss; If there is none, he made the best of this.
                                                                Robert Burns

Three days passed before I showered and dressed. Distance and solitude were again my best friends. I hoped he knew I loved him. I knew he had loved me.

As I thought more about the phone call and the evening prior to it, was it a coincidence that I had my palm read in Skagaströnd? Life could indeed be stranger than fiction, and I couldn't deny the synchronicity between the two events. It was as though I had been forewarned, and from somewhere within, I must now find strength.

I didn't return home for the funeral. I chose instead to do what I thought Mats would approve of, that was, to carry on and face down challenges along the way. And so, despite my pain, I felt motivated. My time was limited in Iceland, and I would continue to make each day count. I had faced obstacles after arriving, both perceived and real. The road conditions and wind had taken me by surprise, and I realised too that changing our geography didn't change our history. The artists complimented me daily on how brave I was to be here alone for so many months, while the locals thought I was simply as mad as a box of frogs, though praised me for writing a book.

Dressed and packed, I looked briefly out the window and saw the waves moving horizontally across the sea. It was blowing a hooley as usual. I had checked the weather app, and it had said a strong breeze in the north, with clouds making way for a little sun after midday. Anyone would tell you if you didn't like the weather in Iceland, wait five minutes. At times, this was true. I took my bags to the car with everything I thought I might need, including my tent which I'd decided wasn't going to stay packed in the bottom of the small wardrobe any longer. I had a vague idea where I was heading. While it was recommended to use camping sites where possible throughout the country, you could camp for one night with a maximum of three tents off-site, though you did need to seek the landowners' permission beforehand.

Literally from walking the short distance from the parked car to the supermarket, no more than six minutes, I had gone from standing upright to being bent double against the wind and trying to shelter my eyes from gravel.

My first solo camp would be an easy drive from Blönduós. That way, if I did run into difficulties, I wasn't too far from home. It was questionable why I would set off with a tent in the middle of winter in Iceland, having not trialled all the equipment first. I had meant to, but having brought the trip forward by a month, there simply hadn't been time. I was not an experienced camper but I have camped several times in Scotland, Wales, and Scandinavia in wintertime, though rarely alone. I had invested in my new tent, knowing it was good quality, warm, light, and easily manageable. It claimed to be worthy of any expedition, and in the Icelandic north, this was indeed one of those! When I had paid the heavy price tag, I deduced that I was going to trust my life in it. A tent kept you warm and dry, but especially in wintertime, that meant safe and even alive. I was ill-prepared for the temperamental weather Iceland threw at me; the high winds especially, though I was confident the Kerlon 1800 material which the outer fabric was made from would stand up to the elements. It was the polar bears I was more afraid of. Mats had desperately wanted me to buy this tent, and I was now glad I had listened. We had even named it together on our way home from our shopping trip. We had named her Bothy, meaning shelter.

Having driven around the Blönduós and Húnavatn area and having walked the gravel roads and paths leading to farms and headlands, I was now familiar with the area and had ear-marked a few possible camping places. I had, through the local grapevine, contacted several landowners and so had gained permission to camp pretty much where I wished, as long as I didn't advertise it and promised to be sensible. I might have exaggerated my level of camping experience a tad and, of course, wouldn't share the information on social media.

There was a great spot, not too far from where I could park the car and then walk the rest of the way carrying what I needed. It wasn't a long hike up to the rock overhang that would provide good shelter from the elements. The spot I'd chosen was flat where a little hardy grass had established itself, just large enough to pitch a tent, and being so high up had no risk of flooding should, by any chance, heavy rain come unexpectedly. One thing I could do was start a fire. My constant hunger and need to grill hotdogs taught me years ago fire was essential. I also loved what we called cook coffee (kok-kaffe) in Sweden, that is, a strong ground coffee that was poured into boiled water and basically left to stew on a low heat with the magic ingredient "time" added, which allowed for the suds to sink to the bottom. I added a small pinch of salt to increase the flavour. There were many different makes, but one I loved was called Lemmel Kaffe. I had brought two packs with me, and it accounted for one of my extreme luxuries. My small cooking stove and aluminium kettle were packed along with my favourite wooden cup which, due to the natural wood grain, had a permanent upside-down smile looking at me when I drank from it. I hadn't noticed that when I bought it on a fly-fishing holiday at Trysil in Norway. It was my grumpy cup.

Although the wind had abated somewhat, I wouldn't describe it as gentle. As I unpacked Bothy, I wondered how I could pitch her with at least one of the entrances (Bothy had two) out of the wind, but once I got on with the task, I found it a doddle. My self-supporting tent with its clip pole system was up and sturdy. I used the extra footprint underneath I'd bought to protect the tent floor from snagging on the rough terrain. I had with me my new Therm-a-Rest and sleeping bag. I was wearing layers of warm clothes and I would be comfortable. Iceland was a safe destination but by no means a safe island. Aside from an extremely rare chance of meeting a polar bear, there were many more real risks I had to consider. Staying warm and dry was essential, and I hadn't brought a heater for inside the tent. The temperature outside wasn't yet below freezing, but the wind chill was. So, with the last of the day's light, I got to work on a fire. I searched, too, in case I found some suitable rocks to heat on the fire which were a great substitute for a heater. Unfortunately, I found nothing, so instead, I walked the half hour back to the car with the aid of my head torch to collect my hot water bottle, which I'd forgotten to carry with me. After that first night, I made a point of finding a few rocks and kept them in the car for future nights "out".

Eventually, my home for the night was built, warm and very cosy, and as the wind whipped the sides of the tent from time to time, I sat back with a hot drink and my knitting and enjoyed the feeling of being "nowhere in particular". It was exactly what I had hoped for when planning my trip, and from that evening, each time I spent a night in my tent, I knitted a square. My plan was that when I returned home, I would stitch them together and make a blanket of memories.

RÉTTIR

Waking early, having slept properly for the first time since Shauna had rung, I crawled out of Bothy to pee. I warmed the remainder of last night's coffee and sat in my doorway to watch three ravens circling overhead. The dark clouds accumulated over the horizon and appeared to be rolling closer by the minute. I checked the weather app, as was my habit now, and it confirmed heavy sleet with strong wind was expected. There was a red weather warning over the immediate area for wind with gusts of more than 40 metres per second. Alarmed, I decided to quickly pack and head back to the car while I could. I worked fast, and yet within the hour, I could barely walk upright. I took each deliberate step carefully and stayed low. After each gust, I marched a few rapid steps, then the wind buffeted me again. I repeated this action, eventually reaching the car. I threw my gear in and fought to shut the boot against the ensuing gale. The drive back was with two hands clasped tight on the steering wheel as gravel bounced hard off the windscreen. The N1 petrol station came into sight, and I breathed out. I turned right and headed straight to the concrete garage block outside the residency. I parked, got out and checked for any damage to the car, but thankfully it was fine. I grabbed my swimsuit, towel, and computer and headed on foot to my favourite café, Ömmukaffé, beside the swimming pool. Inside, it was cosy and warm and the welcome had nothing in common with the weather. I mentioned the extreme driving conditions today, and the locals took a casual look outside, replying with a simple "Ha!" The type of Ha! That meant "What weather?"

Talk turned to farming, and I heard the word Réttir. It was a word I wasn't familiar with but had overheard several times over the past week. I asked Siggy, the boy who served me, what it meant.

"As the daylight shortens and the nights become chilly, sheep and horses are gathered from the high ground and taken back to the farms before the weather closes in. It's serious business but great fun and it ends with a huge party." He grinned. I have heard that Icelandic parties were like those back home in Scotland and felt this was one I shouldn't miss. I decided to work at getting an invitation to one of the Réttir and didn't have to wait too long. That same evening while taking my dip in the hot tub, I struck up a conversation with a local woman who insisted I accompany her and her friends to their Réttir this weekend. Hoorah!

If anyone needed a reason to visit Iceland, this was it! Réttir was one of the country's oldest traditions, not to say one of the most fun, and more surprisingly for Iceland, free to visitors. The farmers, their family and friends, and literally anyone who was interested were invited to come along and help gather in the country's three-quarters of a million or so free-range sheep from their summer grazing in the mountains and valleys. Depending on the farm's location, this could involve a lot of horseback riding or driving. The hard work was rounded off with a good old shindig or Réttaball. Weather dependent, the roundups started early in September and continued through to early October. It was a hugely popular event and no one travelling to Iceland in the autumn should miss a chance at joining in. The name *Réttir* came from the sorting site, which was a large, wheel-shaped pen where the sheep were corralled. Firstly, they were herded into the central section, claimed according to tag numbers, and directed to the corresponding farm's holding pens, then loaded into the lorries for the trip home.

What struck me most was the community spirit. The camaraderie and intertwined relationships of the families were self-evident. Wearing their familiar Lopi Sweaters, all generations from all walks of life came out in force to help. United in their task, laughter filled the air despite the serious business to hand. Just like in the swimming pools, there was no social pretences or hierarchy at the Réttir – they all mucked in. All Icelandic families had roots in farming and fishing, all were part of a community, and as far as I could tell, these national traditions brought out the best in everyone. To a visitor, it looked easy, but when I was offered to try my hand at catching a sheep with very large horns on a slippery surface of mud and shit, I found it wasn't. Steadying the sheep between my thighs, not wishing to grab it by the flesh as I had been warned this could cause damage and even internal bleeding to the sheep, I realised it was rather difficult. I could see why you'd need a few drinks afterwards. Thinking of the popular Icelandic dish of mutton soup with its smell of lanolin and floating cubes of fat, I thought perhaps Auden had a point after all. I had packed a change of clothes for the "ball" but thankfully didn't have time to change before I was herded off to the dance. It had been a lucky miss. Being greeted and welcomed by Lopi-wearing farmers smelling and looking rather shit-splattered, I would have raised an eyebrow or two in the only black dress I had brought with me. Dressing up was not necessary to toast the day's success.

Many people left the Réttir bruised and limping. I was no exception. No one had warned me about the injuries I would incur. I was, however, made to feel one of the gang when sympathetically comparing wounds the following week at the pool.

"Hah," the ladies exclaimed as I tried to maintain naked modesty in the communal shower before putting on my swimsuit, "you did a Réttir. Now you are one of us!" Ha!

I had started to notice Icelanders often used the word "Ha!" said with a distinctive exclamation mark at the end. This was used in many ways, including surprise (with raised eyebrows), amusement (Ha ha), disbelief (Ha!), what did you say (Ha!), you're kidding (Haaa!), I don't believe you (Ha!), told you so (Haa!), you see (Ha!), to name but a few.

Living in the heart of the farming industry in Iceland, there seemed plenty of opportunities to visit farms, get up close and personal with the animals and, for anyone with a strong interest in wool and textiles like the artists that came here, I could see the attraction of spending time in the area, learning and discovering new techniques that they could apply to their craft. I hadn't changed my mind about Blönduós, though. It was still a dump. Artists came and went through the "School" together with daily visits from other artist programmes in the area. One thing they all concurred on was that here in the north, whether you were a weaver, natural dyer, or fibre artist (still not sure exactly what that was), knitter or embroiderer, there was a perfect blend of accessibility to raw materials combined with natural light and energy that inspired and encouraged their work. I understood only some of their techniques and didn't hold any high opinions of much they produced which they called "art". However, even my untrained eye could see the wool on the sheep was of a special quality. The gorgeous natural colours of the sheep must have been perfect for weaving. The horses, too, were not beyond being used in the artistic processes, as I witnessed one day in the residency. It seemed very en vogue to be using a combination of sheep wool, horsehair, and plants to weave with, or for that matter, anything else they could lay their hands on. One or two artists got a little overexcited about heading off to the Kill House to procure some horsehair – a brave move, given they were vegan. I idly wondered if I should warn them that the tail might well still be attached to the horse, but didn't wish to spoil their fun, nor mine. Later that evening the kitchen was quieter than usual, and the two ladies had retired early. The following week, they had dealt with their post-traumatic stress caused by their visit to the Kill House and talked non-stop of the sight of the tail, bloody and raw, that they themselves had been asked to choose from a pile of tails in a large crate and which no amount of washing in the icy river could make clean.

SOUTH COAST

Mid-October, and the buzz and bruises of the Réttir had faded. I was ready to tackle a longer road trip. It felt as though I had been living in this north-western region for months when in fact, it had been only a short six weeks. Gone were my thoughts of leaving early. My living accommodation improved as promised and I now had a desk and chair. I noticed I had become more relaxed, mentally and physically, and for the first time in years had space in my head for new and fresh ideas. I had made progress with my knitting and had even worked a few hours on the Vatnsdœla Tapestry. Since Mats' death, I had taken to lighting a candle as I wrote. Only when it had burned to the quick did I stop writing for the day. On occasion, I had to light two as I had so many thoughts that I dared not sleep until I had noted them down. Finally, my journey felt in full swing, and my dreams were being realised. I had enjoyed several nights camping alone, and my knitted squares were mounting up to prove it, though I doubted neither Ragna nor Jóhanna was particularly impressed by my handiwork.

Blönduós was functional, and I could cope with it for short periods of time, but after that, I felt I would suffocate. Every move I made was being scrutinised in the small town and I started to feel like I was living in a goldfish bowl. The violet-blue peeling paint on the Hotel Blönduós, the rusting corrugated iron roofs, grey buildings with their rotting window frames, shipping containers and the putrid blood from the abattoir that stained the shoreline all irritated me. The footprints of yesterday's walk on the beach might have washed away, but the plastic and aluminium debris were never removed. The Icelandic love of soft drinks bore its environmental scar as much as in any other country, with packaging thrown and blown, never being picked up. Winter was underway, and I had no idea how long the roads would remain passable. Frankly, I didn't care. If I could get away from the town, I'd worry about getting back later. I packed the car, locked my bedroom door, and left the residency without seeing a soul. I drove to the local car mechanic, and with no English, but great efficiency, my studded winter tyres were fitted. My summer tyres were hoisted up a chain to the loft for storage, and I showed him by way of my phone calendar that I'd collect them in February or March. Next stop was the N1 garage to fill up with diesel and windscreen fluid, together with a few bars of Hraun, Yankie, and a packet of Kropp. I smiled each time I bought Kropp as it translated in Swedish to *body*. Kropp were small chocolate drops that looked more like rabbit poo but tasted great. I also told the girls in the garage that they wouldn't see me for a few days, and they reminded me to ring if I got into trouble. I drove out of the garage, turned right, taking the route south as I had a score to settle.

Given that Iceland often topped the world's list of happiest nations, I was feeling positive about my road trip. I was allowing myself unlimited time to drive with no fixed plan, and decided that so far, I had seen only my own Iceland. Now was the time to visit the area my mother, Grace, had been captivated by. To do this, I needed to drive south, past Reykjavik, to the nearest point of the mainland, whereby my nemesis, Surtsey, lay, a few nautical miles out to sea. I knew I wouldn't be allowed to set foot on the Island of Surtsey as it was now protected for research purposes, but perhaps by visiting the south coast, I could understand a little of what had captivated her all those years ago. Finally, the most important journey of all, the one I'd played out in my mind over and over was about to start. All those years ago on her second trip to Iceland, she had travelled from Reykjavik to Heimaey on the Westman Islands and even had a day trip on a small boat with the local mayor of that time to take a closer look at Surtsey. Whether they stepped foot on Surtsey, I couldn't say, but I would get as close as I could.

Situated in the Westman Islands archipelago, Surtsey sat at 63° North. Blönduós, by comparison, was around 65.5°. Surtsey was not a large island. When it reached its peak, it was a mere 2.7 square kilometres. Erosion from the waves was shrinking it constantly, and it was now already diminished in size to about 1.4 square kilometres. This was my chance to confront Surtsey, if for no other reason than to find closure, to say good riddance to the harsh memories it had invoked in me for decades and to put that chapter of my life to bed.

Driving south, the weather was bright and there was no wind. I could count on one hand the days that delivered such beauty and sereneness. A helicopter flew overhead, and I had a mad idea. If there was one possible way I could get closer to seeing Surtsey, it would be by air. By the time I arrived in Borgarnes and stopped at the garage, my mind was made up. I bought two sandwiches, filled my water bottle, and sat down at an American diner-style booth and took out my phone. Firstly, I took a quick look at my Internet banking to see how my funds were looking. Iceland wasn't cheap, in fact, it was way more expensive than I had predicted, and I'd already made a reasonable dent in my budget. That said, other than fuel and food, I bought little in the way of luxuries. My accommodation was already paid for, and if my car didn't suddenly need a repair, I'd finish my trip on budget. Being based where I was, the temptation to shop was non-existent. I also had a small savings account from when my parents had passed away. It was not much, but I had hung on to it for something special. I took the plunge and rang two companies dealing in domestic flights. Taking a fixed wing aeroplane proved too expensive by far, so I opted for a helicopter tour of the south coast. The company confirmed I wouldn't be able to get too close to the Westman Islands but, weather dependent, I might catch a glimpse of Surtsey, which wasn't unheard of on a clear day. I confirmed the booking. I drove the rest of the way to Reykjavik, through the long tunnel and past the garage where I had left behind my handbag. So much had happened since the last time I drove here, and I felt like a different person. I found a reasonably priced, unimpressive guesthouse for the night, and the next morning, followed my GPS and parked at Reykjavik City Airport. The weather was still and cold, with blue skies and unbelievably clear visibility.

"We shall be flying the Reykjanes Peninsular," informed the pilot as he stretched out a hand to me. "Got your camera ready?" He nodded.

"Yes, and a spare." I held up my two camera packs.

"Great, follow me."

"What year were you born?" he turned to ask.

Imagining he must need this information for flight checks I replied, "1966."

He beamed. "Great year. I'm Jon by the way, without an 'h'."

He had my attention.

Pre-flight checks clear and safety belts on, the rotor blades started, and as Jon eased the throttle, we horizontally took off. Seeing the runway, the city, the mountains, and out to sea from the air was breathtaking, and I pinched myself. The magical beauty that was the landscape of Iceland and witnessing the volcanic topography it was famous for was a whole different experience from the air. With my bird's eye view, I swung my camera into action to capture as much as I could while Jon explained passing landmarks. "Hah!" was all I could say in return.

As we flew southeast, tracing the rugged coastline, there was no doubt in the force and power that sculpted this land. I asked if we would see any puffins, but Jon smiled and said they had left for the season. He had never seen a puffin here in the wintertime, so sadly, no he assured me, it would be impossible. Since moving away from the Scottish Isles, I hadn't seen many puffins and hoped this trip might change that. Just like in Iceland, in Scotland, we children went out and searched for baby *Tammie Norrie* or pufflings, who had come out of their nest holes to head towards the sea but had got lost. We gathered them in boxes or carried them in our jumpers down to the beach, and holding them tightly with our thumbs underneath their wings so they could flap, and a few pretend-throws to give them an idea of what was happening, we made a final lurch and launched them into the air for them to flap for dear life and disappear in flight.

As we flew low over lakes and craters, hot springs, and the cliffs and lighthouses of the south coast, Jon asked if I would like to touch down for a brief photo op where the bridge traverses the two continents. Also, a warning light had come on, which needed to be checked. "Nothing to worry about", he assured me, "it's just your door!" I pulled my safety harness tighter though there was no cause for panic. He just wanted to double-check it. Safety was paramount, and I felt in good hands. As we touched down and I reached to open my door, there sitting in the snow, was a puffin. His beak a deeper grey in colour than I remembered, but healthy and plump.

"Well, I'll be damned," said Jon.

"Hah!" I found myself saying like a local. I'm not sure who was more surprised.

Back in flight, and within minutes, the blue sky was darkening, and the good weather was behind us with a low ridge of cloud ahead. Jon advised me regretfully we wouldn't see the Westman Islands today. We talked about Surtsey, and he explained that his father's friend, a geologist, had spent time there over many years studying how life formed on the dead island. The island's first plant, a type of sea rocket, sprouted within the first couple of years. A bird's nest was discovered there in 1970. Today, Surtsey was still restricted from human activity, with only a handful of scientists allowed there annually. There was, however, a thriving seal and bird colony. I had already heard the famous stories of a scientist taking a poop there, and this resulted in a tomato plant growing. It was swiftly removed, as were the potatoes a couple of young men tried to plant while studying there.

Apologising for the Icelandic weather, I thought Jon realised what it meant to me to fly near the coast and see the Islands and suggested I drove south and take the ferry to Heimaey, the only inhabited island in the chain of islands. I had seen my first Atlantic puffin in many years and had a spectacular tour of the craters, the geothermal activity, and the landscape. Icelanders seemed to have great interest in their surroundings and were knowledgeable about the flora and fauna. Everyone was an ornithologist, bryologist, horticulturist, or volcanologist. On our flight back to Reykjavik, I was made to promise I would come back and visit in springtime when the birds made their summer nesting sites. Did I know that Iceland was the breeding home to more than 60 per cent of the Atlantic puffin population? I didn't. Sadly, their numbers were declining. This tour wasn't a standard package but was a perfect introduction to helicopter flying over Iceland, and well worth the hefty price tag it came with.

Without doubt, the day had been memorable, and I couldn't think of a better one. I had seen so much, learned a lot, and for me, the chance to talk to someone who knew of Surtsey and understood why I would make this journey, was validation that I wasn't entirely mad. I had shared some of the stories I remembered being told from childhood, and Jon confirmed many details were correct from that time. Later that evening as I sat in a small corner of an old-fashioned restaurant promising to serve up traditional Icelandic seafood, I thought back to how it must have been in 1963. With a hint of déjà vu, I now understood how our mother could have fallen in love with Iceland. I thought how strong she must have been, to be alone, pregnant, and physically abused, yet determined to survive and even thrive. I drank a glass of expensive, though not very good, wine and silently toasted her. For a moment, I, too, wished our mother had sent her "small islands" to Surtsey, enabling her the freedom she so craved to pursue her own path of work, travel and self-discovery.

I woke up extra early and was eager to start the day. I dressed and made coffee for the road. I left my room key on the hall table as no one else was yet up. It was still dark as I walked through the narrow streets to the unlit car park.

Having flown over the milky-blue lagoon the day before, it hadn't looked nearly as fearful as our bedtime stories portrayed. I was tempted to start my day there but instead decided to drive south and aim for Jökulsárlón and its glacial lagoon without the added company of sticky, silica-clad hair and the all too familiar sulphuric egg smell. Having studied the map, I looked forward to seeing the small villages and towns, black sand beaches, waterfalls, and lava fields which made up the south coast and from where I had heard on good authority that elves would be watching my every move – a change from the residents of Blönduós, I thought. With the population of Iceland being roughly the same as Leicester, I wasn't too worried about the roads being crowded, given that it was October and off-season for most tourists. The evenings, too, were rapidly drawing in, and I would be chasing daylight as I drove.

Before leaving Blönduós, I'd packed the car with everything I could imagine I would need, including the paper map I'd purchased. It felt old-fashioned and familiar to have this open on the passenger seat and a throwback to my early years driving, but I didn't intend to use it to help direct me to where I was going, more a reminder of where I had been. I'd highlighted each route I'd driven so far, and my goal was to have marked all major routes together with as many off-the-beaten-tracks as I could during my escapeation. On the front foot-well, I also had my grab bag with essentials, including swimsuit and towel, Thermos flasks, cameras, lenses, gloves, hat, and chocolate bars, not forgetting mother's handwritten journals which accompanied me everywhere. Bothy lived in the boot of the car together with my camping kitchen, food supplies, water bottles, and my emergency whisky. My first-aid kit was behind the front seat in case I needed it. Driving through the city in the early morning, the traffic was light, and it felt good to see Reykjavik before it woke up. I crisscrossed the narrow streets and eventually arrived down at the old harbour, passed the tourist drag with its bars and cafés, shops and whale-watching boats. The quays were full of brightly painted fishing vessels, ICESAR's training vessel Saebjörg, and the naval coastguard patrol boats Por and Tyr that made up two-thirds of the entire Icelandic fleet. The fish warehouses no longer needed were spruced up and were now restaurants and museums. Driving to the Seltjararenes Peninsula, I checked the clock, parked the car, and opted to walk the rest of the path down to the shoreline and lighthouse. It was low tide and slippery in the morning mist. The air was fresh, and the salty air reminded me of home. Other than a small group of men I took for Greenlandic workers, I met no one. I took an early morning coffee from my flask and a stale Kleina, then headed back to the car and drove to nearby Garðabær, where I wanted to pop into Ikea before driving south.

Ikea wouldn't usually be on my list of must-see places, but I needed a lamp and small table for my room in Blönduós and was tired of waiting for either the staff at the residency or the local charity shop to find me something. I'd also stock up on some Swedish Knäckerbröd, Lingonberry juice, and tinned Sill (herring) from their food department. It would be interesting to see how Iceland had embraced this Swedish institution.

On first impressions, the outer areas were rather shabby and unkempt, which was in keeping with many urban areas I had seen. The car park was busy, and I followed a stream of visitors wearing an impressive array of hand-knitted Lopi sweaters and doubted I'd be mastering those patterns any time soon. How dull I felt in my dark fleece jacket. As I wandered through the store, I noticed it was empty. It was the restaurant they had come to visit. It had the full menu that you'd expect to find in any Ikea, but with an added deli bar – which was new to me. Neatly arranged Smörgåstårtas lined up like soldiers, enormous muffins with garish fondant top hats, and giant cookies the size of side plates. The staff regrettably showed little in the way of customer service, though in their defence, they were preoccupied as they had a problem with the tills. Impatient customers raised their voices, something I hadn't witnessed so far in Iceland. My experience until now had been that locals were almost too polite and well-mannered. I chose a trio of small Scandinavian-inspired open sandwiches and managed to pay, eventually finding a seat sharing a table with a group of young students who told me they lived locally and came here every day for brunch. They had chosen the smoked lamb shank and mashed potato. More than once, they took their plates to the Sauce Bar and helped themselves to the colourful dressings. Lingonsylt, cocktail, ranch, béarnaise, BBQ, honey, and hot wing sauce were all available. Icelanders loved sauce with their food. Next stop was the drinks bar that included a selection of still and carbonated drinks together with the coffee machine offering a range of speciality coffees. This wasn't like any other Ikea I'd visited. Another love the population had was Malt drinks, with Malt and Orange being a winter special. I'd noticed these were popular up north too, and with each can measuring almost half a litre, it was no wonder the hospitals needed well-equipped diabetes units. In my local Ikea back home, I would look forward to a plate of their famous meatballs with mashed potato and gravy, but it was hardly fine dining. Here in Iceland, the girls' brunch of lamb wrapped in bacon, with garden peas, mash and white sauce wouldn't have been out of place in many a gastro pub. Having experienced Ikea Icelandic style, I would be less enthusiastic about future trips to my local store in Sweden. I thanked the girls for their company and, as I left, I counted more than eighty people queuing for the restaurant, and the tables were still full. Ingvar Kamprad knew what he was doing when he opened Ikea worldwide, that's for sure, but surely even he would be surprised how Icelanders had taken it to a whole new level.

Heading out of the store with my flat pack table, lamp and food supplies, I noticed workmen manoeuvring into place a giant goat and realised they, too, must have adopted the tradition of erecting a Gävle Goat for Christmas. This giant straw effigy started as a custom in 1966 in the Swedish town of Gävle and was a huge version of the much smaller *Julbock* or Christmas goat, which during the 1800s was traditionally made from the last sheaf of straw from the year's harvest to ensure a good crop in the coming year. The goat was said to symbolise the worship of the Norse God Thor, who rode a chariot drawn by two goats Tanngrisnir and Tanngnjóstr. In Gävle, it was an annual challenge protecting the goat, or *Gävlebocken* as it was named, from pyromaniacs, and Icelandic arsonists have too apparently adopted the custom, resulting in numerous attempts each year to burn the goat, which was now afforded 24-hour security.

Before being distracted further, I joined the traffic first towards the city and then turned right onto the main road, N1 south. I had given up setting cruise control here as I found it difficult to even reach the speed limits, which were often optimistic given the road conditions. Leaving the urban area behind, I relaxed into the drive and took a quick glance to see if my handbag was where it should be and turned up the music. The road surface was like a patchwork quilt, and for several miles, grit and shingle splashed the car, and I was again worrying about my windscreen. Locals, as usual, were in a hurry to overtake in their huge 4×4s, though I personally was more interested in surviving the journey. The sky was caliginous, and rain threatened, but that wouldn't dampen my spirits. It took less than an hour to reach the town of Selfoss. Concentrating on the road signs, I narrowly missed two sheep wandering along the main road, and as I cussed, I thought they were cutting their Réttir a little fine here in the south.

I had no reason to stop; Selfoss was best described as a functional town servicing local administration and industry. I did, however, take a quick drive through the main streets and found myself on what looked like a recently painted movie set in the middle of the town. I later found out they were in the process of taking traditional buildings from around the country, renovating them and erecting them on one street as a living museum. The result made the rest of the town look even more shabby. I gave Selfoss the benefit of the doubt. Maybe with brighter weather and with businesses open for the summer season, it might look more inviting. Winter didn't do the town any favours. I drove onwards, with noticeably less traffic, through the ever-changing terrain. Farms were dotted at the foot of mountains on my left, while to the right, it was pretty much basalt cliffs and stacks interspersed with barren outlets and wide, dry riverbeds leading out to sea. The flatlands were a canvas of lava, moss, gravel, and even grass which had been planted a little over twenty years ago, a variety that could sustain itself in this harsh lunar landscape. Often beside the remote farms were small churches, and there was evidence too where the original road had been before the N1 was completed in 1974. The only warmth came from the horses that lifted their heads as I drove by. These were not at all like the highly-strung horses I had learned to ride years ago. They were calm, dependable workhorses which in the past had been used not only for farming and personal transport but to deliver mail across the entire island despite the often-extreme conditions.

Tempting as it was, I didn't stop other than for an occasional photograph. There was plenty to occupy me, and as I drove onwards past the rugged black hills, my imagination was piqued as I saw profile after profile of faces staring back at me. This was the strongest case of pareidolia I'd ever had. It was quite surreal to pick out so many images and patterns in one place. It gave me an idea for my contribution to the next artists' exhibition in Blönduós, which I'd been asked to take part in.

In the valley of Eyjafjallajökull lay a well-kept secret. A unique gem hidden away yet only minutes off the N1, it was hard to believe so many thousands of tourists drove past and missed it. An 82 ft concrete open-air swimming pool, built right on the edge of the mountain and fed by a geothermal hot spring that seeped steadily out of the rocks. This pool was built back in 1923 and used for teaching Icelanders to swim. Seljavallalaug wasn't pretty, but it had a raw beauty of its own. A short stop wouldn't hurt.

The pool and facilities were basic, as one would imagine, given its location halfway up a mountain. The changing rooms were sparse but did appear to have had a fresh coat of paint recently. The green algae weed, which despite giving the pool a romantic shade, was slippery and slimy and, on a hygiene note, the actual pool was rank. It was supposedly cleaned once a year, but I was doubtful. In any event, sometimes we must throw caution to the wind and immerse ourselves, literally. Up until this trip, I hadn't bathed outdoors much since childhood but when in Rome ... I took a dip where I could, and there was something genuinely uplifting about it. There was a sense of freedom getting your kit off in nature and stepping into warm waters. I had the pool to myself, and even though it had started to rain, the warm-ish water felt good. I wished I could truly relax, but as usual, I constantly questioned my own limits, capabilities, security even. My aquaphobia wasn't a concern as the pool wasn't deep, but as daylight dimmed and the dull weather helped cast shadows, I felt vulnerable up there alone. I wondered how many cold case murders Iceland had. Mine was the only car and obviously parked for visiting this remote pool, therefore, any attacker could be sure I was semi-naked already. How many ways could I die in Iceland? I stood up in the tepid water and added "murder" to the mental list I had been compiling, which already included by car crash, hypothermia, buried under an avalanche, boiled in a hot spring, blown off a mountain or even choking on a Kropp drop. I hopped out of the pool and threw on my clothes without properly drying myself. From where I had parked, it had taken me less than a quarter of an hour to walk up in the rainy light, the rocky ground uneven and slippery. Now daylight was poor, the rain fell heavier, and I was cold. One wrong foot and I could be in trouble. I took my phone from my bag and transferred it to my top pocket. Mats' warning was never far from my mind.

"If you break your arm, you still need to be able to call for help," he would say. I missed him so much.

He suggested I always try to buy clothes with a top pocket. I was increasingly grateful for his solid advice. I had scoffed more than once at his packing lists – which included two thermos flasks (one for food and one for drinks) – but again knew he was right, as picking my way down the loose scree, I knew I had hot coffee in one flask and hot water in the other which I could pop a couple of hot dogs into to warm up (which I had also bought from Ikea). The thought of warm food lifted my spirits, and I plodded the last twists and turns to the car. This would give me energy to carry on driving. The memory of my dip in the dark pool stayed with me for some days; the algae on my swimsuit lasted forever.

I had learned to understand the Icelandic fascination for pools and hot pots. After all, the geothermic heated water spouted out over the entire island. Swimming was ingrained in the culture, so much so in fact that Icelandic children were not allowed to fully graduate from school today unless they had passed their swimming certificate.

Seljavallalaug was not the oldest pool in Iceland. The "not-so-secret lagoon" at Hverahólmi, a geothermal area near Flúðir, was opened in 1891. From 1909, swimming lessons were given there until 1947, when a new, more suitable pool was built. A few weeks after this south coast drive, I actually visited Flúðir and had by then racked up a fair list of hot pots and pools that I had visited around the country. Flúðir was extremely clean and even had a small natural geyser (Little Geyser) that frequently erupted to the amusement of bathers. I'm 5 ft 6 tall and could only just touch the bottom of the pool. There were a few openings between the boulders that were much deeper and ready to catch me unawares! It was, however, pure joy lying in the blissfully warm waters, 38°–40° Celsius, and far more luxurious than the slimy remote Seljavallalaug. On the downside, Flúðir has become well-known and well-visited, therefore it was unlikely to have the place to oneself, but it was easy to get to and included on many trips of the "golden circle". Around the lagoon, a path has been considerately made for visitors to enjoy. Passing Little Geyser to the Vaðmálahver hot spring, where decades ago, people washed their clothes, then to a rather small but excitable spring named Básahver.

On that first memorable visit to Flúðir, the evening was clear, and I was rewarded with a green aurora dancing overhead. I'd left my camera in the car so simply tilted back my head and gazed in wonder, saving the images to memory. Despite the easy walk on that occasion back to the car, I somehow stumbled and fractured a bone in my foot. It took two hospital visits and several x-rays before they found the fracture and a further course of cortisone once home for the pain to subside and the foot to properly heal.

Thankful to be back in my car, having enjoyed a warm snack and drink, I re-joined the N1 again, accompanied by the steady beat of rain. The locals in Blönduós told me that Vik was the rainiest place in the country, and it didn't disappoint. They hadn't warned me of the wind tunnels. I could see the beach and stack rocks the area was famous for, though I wasn't tempted to step outside. I was sure there'd be future opportunities. I was tempted to find a place to stay but had promised myself a sunrise (optimistic) at Jökulsárlón and decided to press on. The roads up until now had been fine, and save the odd sheep wandering dangerously close to the edge, there was little to look out for. From Vik, I calculated the journey should take me around two hours. In fact, it took me four. The weather app had advised light rain. Statistics said it rarely dipped below zero in the south. All I could say was bollocks to that. The light drizzle was heavy rain, and the rain was now sleet. The wind, at times, blew the sleet sideways, and an ice warning came on my dashboard. The roads fast became slippery even with my winter tyres, and I slowed to a crawl a good part of the last 180 km. Driving in Iceland was for me no different to driving in Sweden. I had gotten used to driving on "the wrong side" years ago. Despite this, many roads were inferior to what I was used to, and my car struggled. It was a two-wheel drive and low to the surface and, therefore, totally inappropriate for this trip. Despite having checked the weather, it took me by surprise, and I drove on into the evening darkness with the temperature falling dramatically. Icelanders would tell you the south coast rarely suffers storms, but by now, I was used to their false forecasting. I barely drove above 50 km per hour, and my palms were wrapped tightly around the wheel. I started to regret my decision not to overnight in Vik, and as I drove on, I wondered several times whether I should turn back.

The previous week, there had been warnings of flooding in the south, with water levels rising in rivers and calls to air caution while driving. I had seen plenty of snow around Mount Esja when I had taken my helicopter flight the day before, and I couldn't see how the south coast could remain exempt. It had obviously snowed in the last day or so despite what any app told me. It seemed that in Iceland, people relied on apps more than common sense, and that included me. My own safety was important but so was my car. It was my lifeline out of here, and I'd be truly stuck without it. I was now both frustrated and worried. I was eager to get to Jökulsárlón, having left the lights and safety of Vik behind me. I wondered, would there be any service stations on the way? I had an idea of places I had hoped to visit, had looked up hikes and trails, black sand beaches and photo stops, but I hadn't given much thought to general service areas. I had everything with me that I thought I might need, together with a host of other equipment I prayed I never would need, including a first aid box, foldable shovel, and a bright florescent orange poncho to be used only in an absolute emergency.

As I passed the town of Kirkjubæjarklaustur (pronounced *kirk-hew-buy-ar-kloy-stir*), I got my first dim glimpse of Vatnajökull, the largest glacier in Europe.

The gradient increased, and the heavy wet snow instantly froze to the windscreen wipers, and visibility became much worse. Thankfully there was barely any traffic on the road. I passed one small car pulled into the verge, which looked like a rental car with people sitting in the back seat though no one in the front. I slowed, in case they needed assistance, but they just waved so I carried on. Literally a mile or so further on, two people were walking. I stopped and asked if they needed help. No, they assured me. They were dressed only in light clothing and were trying to pick up a strong enough Internet connection for a weather report, to decide whether to continue. I let them use my phone, then cautioned them back to their vehicle, where it was parked in the way of oncoming traffic with no lights or warning triangle.

Reaching the Skaftafell National Park, my mood was miserable, so I took a break and went inside the welcoming Visitor Centre. They were closing for the day but assured me to take my time.

"How will the roads be this evening?" I enquired.

"Excellent, they are excellent roads here," the cashier said.

"I mean with the weather, the ice and snow?" I continued.

"Oh, it's nothing."

"I hadn't expected it to snow today," I said.

She looked towards my parked car and said, "Don't you have good tyres?"

"Yes, I do," I responded.

"Hah! You're not used to driving in the winter?"

"Yes, I live in Sweden, and we get winter there." Now I was getting shirty.

"Just don't brake," was her advice.

Excellent, I thought.

"Are you going to see the waterfall?" she asked, obviously hoping to get rid of me.

"No, it's almost dark."

"It's very beautiful," she assured me.

"I'm sure it is, but I will come another day."

"You have a good jacket on you." Again, challenging me.

I thanked her and said I'd come back another time.

"Ha!" she said to my back.

I pulled the car door open against the wind, having forgotten to park it in a helpful direction, and pressed on. The waterfall she had mentioned was Svartifoss, which was a popular tourist attraction accessed from here. Hanging hexagonal basalt columns beneath a 20-metre-tall waterfall, it was located not too far from the car park, less than a 4 km round trip, but there was no way I was doing that walk alone in the dark with wind and sleet. As I indicated to no one and joined the main road, suddenly from nowhere, a pale blue Fiat shot past me doing more than 100 km per hour. It was the group I had offered help to. They obviously decided to press on, too, at breakneck speed. I guessed they were travelling too fast to notice the warning signs.

Despite Route 1 (also known as N1) being called a main road or highway, it included numerous blind turns, crests, single-lane bridges, and a few tunnels. It also had open ditches, cliff edges, and a distinct lack of crash barriers. With the weather deteriorating by the minute and with slushy ice on the road, thankfully the few cars I passed coming from the opposite direction were also driving with care. As well as some experience of driving in such conditions, a healthy dollop of sheer grit was also required to cope with the roads and pitfalls waiting to catch you out. Luckily when one did catch me, I was only driving at around 50 km per hour. The pothole was huge, and there was nothing I could do to avoid it without swerving completely off the icy road. I noticed it at the very last second. The entire bottom of the car creaked and shuddered, and I was sure I had taken a wheel off the car at the very least, trashed the shock absorbers or even shattered the entire suspension. I came to a stop with the rear end of the car half into the ditch. Checking there was no traffic behind me, I put on my hazard lights, got out and took a deep breath. My phone was tucked safely into my top pocket, and I patted it to make sure. It was so dark I could barely see under the car, then remembered the large torch that I had taken as an afterthought during the last moments of packing. I switched it on, and thankfully, it worked. I was shocked when making my way around the car at just how perilous the road was. Despite wearing winter walking boots, it was like a skating rink. The water and slush had frozen and formed a thin translucent film, making the tarmac insanely slippery.

I shone the torch around the car, then got low to check underneath. The wheel itself had some scratches to the alloy, but otherwise, I couldn't see any direct damage. I wondered how long it would be before the tyre suddenly burst or deflated. I had two options, wait and call the emergency services or drive on slowly. I decided to continue driving, stopping at each lay-by to have a look at the wheel. Also, there was now a knocking sound which I suspected could be the exhaust. I pressed on. Maybe it wasn't a life-or-death situation to hit a pothole at low speed and drive into the ditch, but in the cold and dark, even a simple mishap could quickly turn nasty.

Only a couple of hours ago, the weather hadn't been a concern, now it was heavy sleet and the wind had reached around 20 m/s. I knew this, as the last highlighted warning sign I had passed gave the current temperature and wind speed in red. Another reason to stay alert and drive with care!

Anxiously I continued, simply hoping I would reach Jökulsárlón safely and then find a bed for the night. I hadn't pre-booked accommodation, as I was sure I would arrive in daylight and have plenty of options to hand. I was wrong again. Had I booked ahead, they might have notified the emergency services if I didn't arrive. I could have asked them to do that as I was travelling alone. My tent was more than adequate for an overnight stop, but that, too, meant no one knew where I was. Finally, a suspension bridge came into sight (hengibru) and as I drove over the outfall river from the Jökulsárlón lagoon, I caught sight of my first icebergs to my left. This was the foot of the Vatnajökull National Park and the runoff from the Breiðamerkurjökull glacier. I had made it. I turned left into the car park, where there were no cars, and the small buildings with signs for café, information centre, and tours were all in darkness. What now? I took another opportunity to check under the car, it looked no different, and in the few minutes I was outside without my hat or gloves on, I was frozen. I took out my phone and searched for places to stay. Thankfully throughout the trip, I had so far never lost Internet connection. I found a room just a short drive onwards and booked it immediately. I'd stay the night there and then come back to the lagoon at sunrise. I wouldn't need to get up too early: the sun didn't. With little to see in the dark, I drove the few kilometres to my accommodation, checked in, and decided to put my worries aside for the evening. I would tackle whatever tomorrow threw at me, tomorrow. I was delighted to find a warm, comfortable room with a shared but immaculate bathroom and even a kitchen with a coffee machine. It was more like a holiday village than a hotel or guesthouse, made up of various bungalows. As far as I could make out, I was the only person in my bungalow, which was surprising given the number of cars in the car park. I showered and got into bed. As I turned off the bedside lamp, I knew that I would sleep. Since arriving in Iceland, sleep came more naturally, and my dreams were more peaceful.

Morning came, and as I reminded myself where I was, I felt a surge of excitement course through me. The reality that I was here, beside the Glacial Lagoon that I had seen pictures of for so many years, felt like I was an extra in a David Attenborough documentary. Not so many miles out to sea lay the Westman Islands and Surtsey. I'd passed the signs yesterday. I toyed with which direction I would take later in the day. I could carry on and do the entire circuit of the ring road, or I could head back the way I had come. From the north, the route I had driven to get here was around 600 km. If I continued the southeast coast, it was nearer 750 km. I knew if I continued along the south coast, eventually I would have to cross some steep mountain passes, and given the change in the weather, I wondered whether my car was up to the job. I could drive from Blönduós clockwise, weather depending, at a later date. I had also toyed with the idea of taking the ferry over to the Westman Islands, as far south in Iceland as I could be, as near to Surtsey as possible, just as Jon, the pilot in Reykjavik, had recommended. My telephone reminded me to light a candle. Not to write by but in memoriam of my mother's birthday. I'd completely forgotten about it but took it as the sign I needed to head west and catch the ferry – but not before I had kept a date with diamond beach.

In the morning light, and with better weather, the views were very different. To my left, I could now see right out to sea. The wind was blowing but not as violently as yesterday. Driving conditions weren't great, as a light coating of early morning snow had blown in light drifts across the two-lane highway. The car park at the glacial lagoon was almost empty, but I imagined in an hour or two, it would be busier.

Words couldn't truly explain this natural marvel. Breathtaking, mighty, immense, stunning. They meant nothing compared with what you did see. I had imagined it to be a still and silent place; it wasn't. Formed by melted glacial water from the Breiðamerkurjökull glacier, the lagoon was home to active hunks of fantastic ice sculptures making their way slowly out to sea. Traffic on the lagoon was busy, and as the ice boulders crashed into one another, the cracks were explosive. A seal popped its head up and darted underwater again, but not before I got a chance to get a quick photo. Not quite knowing where to start, I headed up the lagoon, climbing the series of viewing points for a panoramic view. It was windy, so I hung on to my tripod with one hand the entire time. Previous experience had taught me never to let go.

On a clear, cold, sunny winter's day, this spectacle would be incredible, but even in the morning dawn, with clouds overhead, it was a vision. I hoped I would have the opportunity to come back again and do some serious photography, but today I was happy to enjoy the moment and catalogue the memory. I enjoyed the solitude of the morning and slowly walked the area around the lagoon and down to the black sand beach where half-buried ice blocks of all colours, shapes and sizes lay like beached whales. Over time, some had collected ash and dirt as they froze. Others were crystal clear or milky opaque like Lalique sculptures. Everywhere I turned, I was greeted with a new piece of art. Displays in the finest museums and galleries throughout the world could not compare to the exhibition on show here. The wonder of it all was that, by evening, it would look different, and again tomorrow and the next day.

Waves with a strong current crashed up onto the beach and surrounded the ice with foam. I thought about my bathing suit in the car, then reminded myself of the dangers of these beaches and the strong current and had no intention of swimming. Yet foolishly, I couldn't resist a toe dip and so set my camera to timer and took a very quick walk into the sea, the force of the water taking me waist deep instantly. This was not at all recommended. Afterwards, I realised it was plain stupidity.

I'd enjoyed the best of the morning, and with a wave of tourists descending on the beach, I walked across the road back to the car and drove onwards with a final glance over the bridge up into the lagoon. I promised I would come back soon, and I did so, several times.

HOME ISLAND

The drive from Jökulsárlón in the direction I had come from the day before was far more enjoyable in daylight and certainly quicker, even with several stops for photographs along the way. A uniqueness of this coastal route was the proximity of the mountains to the ocean. You could see where the sea level had receded, land filled with volcanic ash, and even an island that once stood in the sea was now literally surrounded by land rather than water. It was a rare privilege to see this so close to a main road. The dramatic scenery made this one of the most spectacular driving routes in the world. I was surprised it wasn't busier, though grateful too as I was sure there would have been a far higher accident rate.

Jon the pilot had told me to head for Þorlákshöfn from where I could catch the ferry to the Westman Islands. In wintertime and in bad weather, the ferry took this longer route with a crossing time of around 2 hours 45 min. In summertime, it left from Landeyjahöfn and took only around 35 minutes. I checked the online timetable and there were two crossings, one in the morning, which I would miss and one in the early afternoon. I'd aim for that. The islands weren't exactly a tourist magnet in the winter months, and even the puffin population had now migrated. I was surprised the ferry operated twice a day. Only one island in the group was inhabited, named Heimaey or Home Island, home to around 4200 people and countless birds. This was the island Grace had travelled to almost six decades earlier and I was excited to be following in her footsteps. This was my opportunity to get as close to Surtsey as possible. I hadn't planned it on Grace's birthday; it just worked out that way.

I arrived in Þorlákshöfn. It was at first sight as appealing as Blönduós had been, though quickly, I realised it was worse. The town itself had the same low functional buildings, the same rusting containers and debris, the same illusion of being left behind long ago. This was low season, and I forgave businesses being closed, but there really was nothing to like about the place. I located the ferry slip without difficulty and checked the information board. My heart sank. Apparently, there was no afternoon crossing. I'd have to wait until tomorrow morning.

I scanned the town with its familiar collection of green, blue, and red roofs. Beyond the harbour with its industrial units and fish processing plant, lay the residential area, the usual collection of bungalows, houses, a church, and swimming pool. I couldn't face a night there even if a guest house were to be found, so I drove out of town past the long black powder beaches and found a very out-of-the-way cove; I could pitch Bothy for the night. I had the remainder of my Ikea hotdogs, herring and bread, and plenty of chocolate, together with several half-full water bottles in the car. Despite the car heater, I had felt cold the entire day, probably due to my dip in the sea at Jökulsárlón. My entire body ached, and I craved a hot meal. Food was important, a necessity to replace energy and ward off the cold, and I knew I'd been eating too little. My weight had dropped since starting this trip. I took long walks with my daypack, and even when I used the car, I was active between stops. I had noticed recently not only that I had lost weight but also that many of my menopause symptoms had all but disappeared. I no longer suffered from hot flushes (I was freezing cold most of the time), I had better muscle definition than I had had in years, I hadn't offended anyone with my mood swings (probably as I had barely spoken to anyone), I slept well, and as for the more southward side effects, I couldn't comment as I still had only used my crockery and cutlery for one. I didn't think I had lost my libido and there was still time on this trip to check that out should an opportunity arise.

The town of Þorlákshöfn might not offer an abundance of restaurants, but I had enough food with me, and as soon as I had made Bothy comfortable for the night, I warmed up some food and ate dinner.

Packing up my camping stove, I still felt cold. I knew the town had one thing I longed for, so I made my way back up the headland, got in the car and drove the short distance. I parked in front of the swimming pool and was happy to see they were open until late. I quickly paid, showered, changed into my worn-out algae-stained swimsuit, and eased into the warm water. This was a luxury I would miss when my time came to leave Iceland. One by one, locals arrived, and they, too, eased into the pool, relaxing away their daily stress. They made polite conversation, and I was surprised to learn there was an expansive building project on the go that would make this town the largest residential neighbourhood outside of Reykjavik. I'd never have guessed that so many people wanted to live here but kept my opinion to myself. Plans, too, were underway to attract more cruise ships due to its deep harbour. As I crawled into my sleeping bag later that evening with my wooden mug of tea and my knitting, I treated myself to a tot of Bowmore from my miniatures collection. "Happy Birthday," I toasted Grace, feeling sure she would have approved of my choice of accommodation. On the ferry the following morning, an older gentleman who had waved me aboard came and chatted and asked where I was staying. His sister had a spare room to rent if I was interested, as her eldest son was now studying in Denmark. I kindly declined, explaining this was a short visit and that I'd been comfortable camping overnight at a secluded spot further along the coast. He asked questions about where exactly I had stayed and then declared my secret nook was, in fact, a well-known beach used for filming several movies, including *Flags of Our Fathers*. He also asked if I'd enjoyed my evening at the pool. I started to feel there weren't many secrets left in Iceland.

I flicked through the dog-eared tourist brochure on board, which gave a little insight into the islands. Two historical markers seemed to have made this island what it was today. The Turkish Invasion of 1627, and during 1973, a huge volcanic eruption which rocked its foundations, resulting in a total evacuation of the population. Miraculously, everyone survived and many made their home in Þorlákshöfn, especially those relying on fishing for their income, given its natural, deep harbour. There was no mention of Surtsey. I drove off the boat and made the most of the few hours on shore. I'd bought a return ticket as the older man had said the weather was not looking promising for the coming days and the ferries could be cancelled at short notice. I answered confidently that the sky looked bright, and the weather seemed much better than the last few days. Ha! was all I got in return.

With the winter lights reflecting off the fine dusting of fresh snow, Heimaey looked like a Christmas card. I drove around the small town of Vestmannaeyjar and stopped by a cemetery with multi-coloured crosses illuminating the graves. I thought about how dark our island cemetery was back on Orkney. This beautiful tradition of remembrance lights was popular in Finland and Poland too where the winters were dark, and whilst the colourful electric lights used here looked a tad tacky, the idea was a good one. In Sweden throughout winter, they lit predominantly white candles or LED look-a-likes that made for a beautiful spectacle from All Saints Day onwards. Vestmannaeyjar was compact and neat with cafés, museums, a puffin hospital, and a Beluga whale sanctuary. Most attractions were closed though a couple of museums did look open. I chose first to drive out of town and stretch my legs, thinking perhaps I would have time to visit them later. I'd been sitting in the car for days and now needed to move. I parked and took a well-worn path up Mount Eldfell, the volcano that resulted in the mass evacuation in the 1970s and admired the view. The hike was easy, and I noticed there was no snow here. I guessed the warmth within the ground must have been enough to stop it from settling. I sat down, and indeed, the ground was warm. I kept a candle and matchbox in a small tin in my daypack. I took it out and placed it in my empty coffee cup and lit it. I offered a silent prayer of forgiveness for what I had been dealt during my formative years and to a mother who knew no better. Surtsey was just a landmass. Iceland was just a country. My hanging on to the past had driven me this far, even pushed me beyond my comfortable limits and had brought me to Iceland. It had for decades weighed down on my shoulders like a lead shawl, and I wanted to be free of it. It was a negative energy I had tried to harness and use positively, but, like the bad relationships in my life, I was also letting this one go. I knew that the strength to follow dreams came from within. This trip had proved to me that I could follow my heart, use my head, and make choices for myself. I couldn't see Surtsey from my elevated seat but knew it was out there, just a lump of rock and home to seabirds and seals, nothing more. I took my telephone from my top pocket and rang each of my sisters in turn. They exploded into exaggerated laughter when I told them I had taken myself to Surtsey or as near as I could. We reminisced for a few minutes, and then, with all humour gone, each had asked how I was coping after the loss of Mats. As my candle flickered, tears of joy turned to tears of sadness.

"Forget fucking Surtsey," said one.

"Come home safely," said the other.

I promised I would do both, eventually.

Now, however, too much coffee and the whisky from last night rumbled, and I suddenly needed the loo. There were no public toilets in sight, and I hadn't remembered even seeing one. I panicked, wondering where I could hide and barely had a chance to do so before I was forced to dive behind a small bush and answer my call of nature. As I squatted, I humorously thought this was a fitting final curtsey to Surtsey.

The few hours I had spent on Heimaey were more than enough to see what there was on offer on a winter's day. I still had an hour before I needed to catch the ferry, so popped into the Eldheimar museum. They had a section on Surtsey, albeit with nothing I hadn't already seen or heard. The woman who took full payment for the brief visit wasn't keen on talking to me about it either. Her mobile was more interesting. As I boarded the ferry, I felt somehow lighter than when I'd arrived hours before. Events made us who we were, and my brief visit to this small island taught me one thing. I could escape to a rocky outcrop in the north Atlantic, but the pull of love and friendship would prevail over distant solitude. I'd stay the planned months here, though for the first time, started to look forward to going home for the right reasons.

The ferry rolled into the small harbour, and I hastily drove off and left Þorlákshöfn behind me. Many times, I had thought how the journey mattered more than the destination. I continually searched out towns, villages, landmarks, and even historical sites, looking for what was interesting, only to find the opposite. This wasn't down to unrealistic expectations, but more so that not everywhere in Iceland was worthy of a superlative or was even nice. Talking of which, I was now in my car and heading once again back to Blönduós.

The speed limit might have been 90 km per hour, but I barely reached that. In fact, for more than half of my entire 1300 km round trip to the south coast and back, I hadn't risked a speeding fine. When I met other tourists, the most frequent question asked of me was how the roads were, could I recommend any places a little off the beaten track, and could the beaten track be conquered in winter in a regular car?

Each time my answer was the same. Firstly, all the sound advice in the world couldn't help you if you weren't honest with yourself. Know your driving capabilities and experience. If you felt you had no fear whatsoever on these roads but could answer honestly that you haven't driven consistently in such conditions, then you were fooling yourself and were a danger to other road users. If you had some measure of concern and allowed time to research changeable conditions on the way, had taken them into account and understood that you would have to adjust your own driving to suit the conditions, then you were halfway to a safe and enjoyable experience. If you were terrified of the prospect, then you should probably just accept the fact that driving in Iceland wasn't for everyone. I constantly checked (www.Vegagerðin.is) in English, which gave a live update of all the roads in the country. It was colour-coded and showed layers of information. Webcams were placed along major routes, particularly at dangerous points or the high grounds so you could view live snowfall and drifting, flooding, and accidents. You could even monitor the tractors' progress in clearing roads and see route closures. The app advised on a range of driving conditions from easily passable, spots of ice, slippery, extremely slippery, wet snow, difficult driving, very difficult driving, and road impassable or road closed. You could choose various additional options to show actual wind speeds or the number of cars that have used the road during a specific period. It wasn't just winter driving that caused problems. Fog, sandstorms, and water levels could all impact safe driving at any time of the year. Despite this multitude of information, I reminded myself to rely on common sense. One person's idea of difficult driving wasn't another's. Having spent weeks in Iceland translating the weather forecasts, there was a trend.

"Spots of ice" meant the road was at least half covered in ice you probably couldn't see, but it was there, along with a fair amount you could see, which itself was scary. You'd hang on tighter to your steering wheel and not dare blink.

"Slippery" meant driving on a wet ice rink. This white-knuckle experience left you exhausted when you could finally come to a halt, albeit slowly, because if you touched your brakes, you'd most certainly be off the road.

"Difficult driving" was little less than potential suicide.

"Light breeze" meant hold on to your hat, or it would blow away. Don't even be tempted to bring an umbrella to Iceland.

"Fresh breeze" meant a light breeze with an arctic chill driving the temperature down drastically. This resulted in painful chill-burn to your face and other exposed extremities.

"Gusts of wind" meant if you could stand up, you were doing well, but to stand up with your face in the wind and breathe at the same time was impossible. Additionally, the paintwork on your car was history if you found yourself driving anywhere near loose gravel or sand.

A misconception about driving in Iceland was to think if you stick to the ring road, it was safe. I, too, fell for that one, though quickly learned otherwise. It, too, could become impassable quickly, and on one occasion, I witnessed the entire ring road, all 2250 km of it, closed at the same time due to high winds. Almost on a weekly basis during winter, sections of the ring road were closed for hours or even days. There was constant talk of the weather being unpredictable, but what news reports failed to mention was that Iceland was a "one road in, one road out" type of country. If the main road was closed, where were you going to go as an alternative? The answer was you weren't. As the day approaches for your flight home, be careful. Was seeing that one extra waterfall or mud pot worth becoming marooned for, unable to reach the airport?

I tried to listen to my own advice. The east coast was particularly concerning me, especially between the towns of Mývatn, Egilsstaðir, and Seyðisfjörður. That was the road I first drove when arriving by ferry. The mountain pass hadn't been easy in late summer, and as the weeks passed and my return ferry was booked some weeks ahead, I obsessively checked online and tallied up that on more than 50 per cent of the days, it was impossible to reach Seyðisfjörður from Blönduós. There was no refund policy on the ferry ticket for a no-show, and there was no extra insurance which I could take that covered me for inclement weather. This was nail-biting when my date of departure approached, as I watched the weather worsen and major sections of roads close. In fact, when I could bear it no longer, I headed off days earlier than necessary to give myself a fair chance of arriving on time. It worked! Literally a couple of hours after I arrived at Seyðisfjörður, the road closed behind me, cutting the town off for several days.

The wind was the most dangerous element of all here. Having travelled to a heck of a lot of places that claimed to be windy, including New Zealand, Australia, and parts of the US, I could say with authority, forget it. While certain places in the world occasionally cooked up a storm, Iceland was on a constant slow boil. Unless you had experienced Iceland full-on, you haven't felt the true force of wind, which could and did arrive without warning. It was an assault on your senses, a slap to your face and body and could easily pick up your car and throw it into a cycle of rage as fast as you could say "fresh breeze". On an extremely sombre note, a dear lady who befriended me in Blönduós, told me the story of her son, a healthy, happy young man with his entire life ahead of him. He headed off to Reykjavik on the bus one day and came home in a coffin. The wind over the mountain pass was evil, the same pass I had struggled with several times. On that fateful day, the wind reached speeds of outrageous calculation and the bus he was travelling on, a regular bus service, was lifted in the air, rolled several times, and dumped to the ground. Families were robbed of loved ones and suffered life changing losses that could never be replaced. I had been accused of obsessing over the weather and couldn't argue with that. It dominated my trip and yet even then I was distinctly unprepared at times for what was thrown at me.

Three weeks after my initial south coast road trip, I again drove from north to east, then south to return to the glacial lagoon at Jökulsárlón. I had been fascinated by the place and couldn't wait to return. I hired a 4×4 land cruiser on the good advice of the search and rescue gang who used the pool in Blönduós. It was by now mid-November, and as I hugged the coast between Höfn and Breiðdalsvík, the wind speed crept up more forcefully than anyone predicted, and gusts registered an alarming 51 m/s. The mountain pass with the ocean on my left was steep and intimidating even if visually stunning. With one almighty gust, the Toyota, almost two tonnes in weight, lifted sideways onto two wheels. I was extremely close to being blown completely off the road. A parked tour jeep yards ahead witnessed what happened. The driver flagged me to stop and checked I was ok. I said I thought I felt the car lift, and he confirmed I had only two wheels on the road for some seconds. I was shaken but otherwise fine, and after a cup of tea from my thermos, drove steadily throughout the remainder of the journey. That event taught me that I was right to obsess about the weather in Iceland and unafraid to drive cautiously in bad weather no matter how many would-be rally pros were speeding recklessly around me.

The blame for road traffic accidents in Iceland couldn't only lie at the feet of tourists. Local drivers constantly underestimated the weather. I couldn't understand why the Icelandic media consistently downplayed the weather. Reading the news online daily, with frequent regularity, it informed us that temperatures didn't drop far below zero in Iceland, and that storms, ice, and snow were rare. On a regular basis, I woke to snowfall, high winds, and gales, with even a double hurricane or two recorded. This "freak" weather caused chaos on and off the roads. An example of their reporting went like this:

Warning, river levels exceptionally high; Cold Snap! − 18° by Monday; Storm warning for East Iceland; Storm predicted on highland roads; Big freeze for this weekend; Iceland in grips of cold snap; Severe winter weather on way; No end in sight for winter weather; Storm warning for Friday; Ten ways to survive a "Snowpocalypse";

Severe weather closes stretch of Ring Road; Warning – people in Grafarholt and Grafarvogur should stay indoors; Weather warning – don't travel today; Hurricane force winds predicted for tomorrow; National Police declare state of emergency today due to hurricane force winds; Storm – pick up all children before 3 p.m.; 600 km of roads already closed; State of alert in South Iceland raised to hazard level; Power cuts very likely; Evening flights postponed; Weather goes crazy in Reykjavik; Armoured vehicle to the rescue; Weekend temperatures plunge; Weather never seen before; Storm warning for the South tomorrow; Storm warning for highlands; Storm chaos in Capital.

That didn't sound like unusual or freak weather to me!

Despite or perhaps because of the dominant weather, my journey had quickly become the peregrination I'd dreamed of. When meeting and chatting with other tourists, I realised just how much more I was getting out of the trip than the average visitor. Having given myself time and the luxury of a car, I was making a daily acquaintance with the real Iceland. My slow pace of travel allowed me to find the surprising and superb, as well as the not-so-impressive. Icelanders were survivors by nature, and that survival has meant throwing open their country to tourists. That didn't mean everyone liked the idea, and often in smaller towns or villages, their behaviour towards tourists or visitors was distinctly cold, bordering on rude. Regularly when I arrived for the first time in a village and smiled a hello, I was met with a cold gaze or a turned back. If I had the opportunity to speak, I found myself quickly explaining I was from Scotland, just across the water, hoping that would somehow help. On occasion, given the fishing heritage and history here, this worked as an ice breaker though just as often, the canny elder folk with lightning speed mentioned the Cod Wars, the only war Iceland has been active in and yes, it was about fish. Iceland was victorious in achieving its aims. Ha! After some weeks in Blönduós, I had nurtured a reasonable communication with local people, but as soon as I drove a few miles outside of the town, I was again simply a tourist to be tolerated, to earn money from, and preferably, to say goodbye to as soon as possible, or so it felt. Travelling alone was, at times, a lonely experience. I found myself saying, "Sorry, I'm alone" or "Just the one" when taking a seat in a café or buying a ticket. Days could pass without a proper conversation. I kept myself entertained with the weather, the roads, and the ever-changing scenery.

With my luxurious 4×4 hire car, I accessed many remote spots. The risk was that when I did get stuck, I was more out of the way and so less likely to find help. The highlands were covered in snow, and even the east and south coasts were getting some serious winter weather. I took my time and stopped when I found anywhere that looked interesting, was picturesque, or had a loo. I also stopped whenever I saw graffiti. I'd never held much interest in it before, but it was impossible to travel around Iceland without noticing it everywhere. It met you in the most unexpected of places and was far from vandalism. In the uber-trendy capital Reykjavik, it was almost expected to see art everywhere, and Street Art or Urban Art enthusiasts couldn't fail to be impressed. When walking around small, out-of-the-way villages or towns it was less expected, but again, Iceland didn't disappoint. From the south coast to the east and even far north, time and time again, I was rewarded with magnificent murals. Some were basic or crude throw-ups and stencils traced back to artists who have been in residence in these small towns, others were impressive wall murals by international artists such as Guido van Helten and even one rumoured to be the work of Banksy. My personal favourite was in the northeast town of Reyðarfjörður. I parked the Land Cruiser and walked around the small town, hoping to find a clean loo before making myself some coffee. Freezing fog hung in the air, and the surrounding mountains gave the place a dark feel. There was little of beauty there, the small town being home to an aluminium smelter and little else.

Surprisingly as I walked around a shabby grey building, I was met with a beautiful graffiti mural of an Inuit hunter casting his harpoon, painted onto a huge sliding door of an old warehouse. To its right on the wall, an angry arctic wolf. I raced back to the car for my camera, and as luck would have it, met a fisherman. He directed me to the harbour offices, and they generously allowed me to use their bathroom. They thanked me for asking, as apparently many tourists simply squat and go at the side of the roads. I didn't tell him I'd been forced to do that a few times too.

Another unique outdoor art form was the knitting-graffiti or yarn bombing! Lampposts, bicycles, and signs were adorned with knitted or crocheted coverings right down to small pebbles used as doorstops. Especially popular in Blönduós, these colourful additions broke up the grey of winter, and I often wondered why more countries hadn't adopted this tradition too.

THE LIGHTS

Having driven from Blönduós east and then south, hugging the coast as much as the roads allowed, I had traversed the entire north, east and south coast routes. Now I headed northwest, and as I passed the outskirts of Reykjavik – the chlamydia capital of Europe – I felt ready to once again head north and back to Blönduós – the diabetes capital of Iceland – but not before I had made a detour over the Snæfellsnes area to the town of Grundarfjörður. This was my second attempt to drive this peninsula. The first saw my Volvo and I stranded for two days, paying the equivalent price of a five-star hotel to camp with Bothy in a farmyard, which I chose to forget the moment I left. Little did I know that it wasn't the last time I would be stranded in the area pleading for anyone with a 4×4 to come to the rescue. I felt proud thinking of the miles I had covered and congratulated myself on having highlighted the entire circumference of Iceland on my map, with hundreds of stops and detours along the way. Having the land cruiser had opened roads impassable to my Volvo, and this most recent road trip had allowed me to see more of the very rugged interior. Had I been staying in Iceland any longer, I would have been forced to trade in my car and buy a 4×4 and pay whatever exorbitant rate the formidable customs lady might have demanded. But in fairness, my Volvo hadn't let me down. It had its limits, but it was sturdy, well-insulated from the wind, and felt safe. If there was a road of sorts, it coped well, even if it did grind along, ploughing the mud or snow as it went. I had two days left before I had to return the hired Toyota. I'd planned to visit Snaefellness, then drive over to Húsafell to keep a special appointment. I'd been offered a chance to go deep inside a glacier, and that was a date I was not about to miss.

As time passed, so did the drizzle, and on reaching Borgarnes, the sky was clear. As I drove over the long bridge, the bay was bathed in an apricot glow of northern lowlight. I refuelled the Toyota and checked the windscreen fluid. Although it was getting late and already dark, I pressed on, thankful for decent driving conditions and looking forward to reaching Grundarfjörður, hopefully in time to catch the restaurant on the harbour open, as it came highly recommended.

Even with the 4×4, I had checked the weather reports as usual. What I hadn't checked was the aurora forecast. As I drove on, enjoying the peaceful evening, suddenly, I noticed clouds forming ahead of me, becoming stronger and gaining colour. The moonlight was dim, a sliver of moon hanging low in the sky.

My first experience of an aurora was as a child. On rare occasions on our islands, we saw a greenish hue in the dark winter sky. Later, when travelling in Lapland, I witnessed my first true colourful, blood-red spectacle. I had been ice fishing on New Year's Day, a short distance outside of the town of Rovaniemi. Since then, I'd yearned to see more. Since arriving in Seyðisfjörður on the first day of September, the lights had been regularly active though they had dwindled over the last two weeks to almost nothing. Most nights, I'd spend cold hours waiting, searching for the elusive lights, even though the forecasts weren't good. The last time I had seen them there had been an incredible solar storm, and that brought with it extremely high aurora activity: an 8 out of a possible 9 was forecast on the Kp Index. This resulted in me racing out of town to a high point, pitching Bothy and boiling water for my thermos. Then I set up my tripod and both cameras with remote controls, keeping a night vision torch to hand. That way, I could stay warm and comfortable for the long night ahead. Every aurora sighting was dramatic, with no two shows the same. However, since that night, the heavenly scene had declined to show up.

While concentrating on the road ahead, I hadn't been thinking about seeing the lights, so when they appeared, the spectacle was even more magical. As I drove onwards towards Grundarfjörður, the temperature on the dashboard read minus 3°, the stars and planets twinkled high in the sky while the most glorious green aurora lit the horizon. The defined contours and colour of the spectacle ahead forced me to drive slowly. There was no other traffic around. The aurora kept me company the entire hour-and-a-half journey on Road 56 over the mountains, the area most famous for the earliest settlers to Iceland, past the towns of Stykkishólmur and Kolgrafarfjördur – which famously hit the news when 52,000 tons of herring washed up dead in the fjord, apparently due to suffocation. I drove into the small town of Grundarfjörður and was glad to park up in front of the harbour, get out and stretch my legs and gaze up at the silhouette of mount Kirkjufell, probably the most photographed mountain in the entire country. I grabbed my overnight bag and handbag, locked the car, and walked to the accommodation I had booked right on the harbour, accompanied by the smell of fish.

A common question other foreigners I met asked me was how they could see an aurora. I was always truthful and tried to explain the perfect conditions regarding weather, moonlight, and location. That said, no amount of careful planning would guarantee you success, but there were factors you could consider, helping your odds. The aurora borealis was a natural phenomenon and was visible when radiation from the sun entered Earth's atmosphere and clashed and collided violently with gas atoms in the atmosphere. The lights didn't "come out" at night. The conditions simply made it possible for the naked eye to see them. They were also there through the daytime. The colours of the aurora arose from various Earth gases. Green, the most common colour, came from oxygen. Blue and red were rarer and came from hydrogen and nitrogen.

There was a luck element to seeing "the lights", and you couldn't force it. Firstly, plan your journey around the place and not the event, otherwise you raised the odds greatly of going home disappointed. There were many wonderful things to see and do in Iceland, and to witness an impressive aurora was the icing on the cake. If you asked the same question of an Icelander, they would tell you to walk on a glacier, spot humpback and orca whales, take a swim in a warm pool, eat the freshest fish, drink the purest water, and breathe in the cleanest air imaginable, and let that be enough. If you were fortunate enough to see the lights in all their magical glory, take that with you as a cherished memory – which has likely been prompted by elf karma! If you wished to help the elves in their quest, get as far away from ambient light as possible. That wasn't to say you wouldn't see an aurora in Reykjavik, but the darker the sky, the better the show. Opt for a lunar phase with as little moonlight as possible. Clouds would distort the view as the aurora was above the clouds, so a clear night was optimal. The best time of year was between September and late March. Often early in the evenings, the lights could shine brightly then disappear for a few hours, reappearing late into the night. It was a misconception that it had to be cold. The cold itself didn't help, though if the weather was colder, that often gave rise to a clearer night sky.

I was no expert but spent months learning from my own mistakes. I became a fan of the Icelandic weather site www.vedur.is, as not only did it provide an accurate weather forecast but an aurora forecast as well. It included the Kp (global geomagnetic index). The scale ranged from 0 to 9. Depending on which latitude you were at, you could use this as a guide. In Iceland, anything above 4 meant a good chance of seeing the lights, but I've had numerous reasonable sightings with a lower number. On the other hand, one evening I had an 8, which was classified as "severe" in a good way. The lights were fantastically bright, although this wasn't the most colourful or impressive display I've seen.

In the moments of excitement before running outside to see the lights, I learned quickly to apply a little common sense. Dressing warmly ensured you wouldn't give up after five minutes. The lights came and went and came again, developing in strength and colour. They often played a little hard to get, teasing the spectator and then suddenly revealing their full glory and dance across the sky. Imagine missing that because your hands and feet were cold. Wherever I was, be it in Blönduós, in a guest house, or in my tent, I left my clothes ready to jump into. I also set my camera on the tripod with the appropriate settings. I found it almost impossible to fix everything outside in the dark with cold hands. Mats had taught me that to fail to plan was to plan to fail, so I went to bed with clothes and equipment ready and my green light headtorch in order not to destroy my night vision. I also kept a packet of hand warmer sachets ready in my pocket in case it was colder than I expected. It had been a painful learning curve on many an evening before I realised for myself the importance of planning ahead.

Photographing the lights was a real challenge, yet with the help of Nikon's homepage and articles, along with trial and error, I got better over time. To be honest, my iPhone took great snapshots, but nothing compared with my Nikon D850 camera and lenses. I used both 14mm–24mm and 24mm–70mm. I also used a tripod. I made the expensive mistake one evening of letting go of my tripod in the wind, only for a moment, assuming it was anchored well. It wasn't, and I smashed both my older Nikon camera and a lens. What you see with your naked eye was hugely increased through your camera lens. I took hundreds of pictures of grey smears of cloud only to upload them and see swirling cones of yellow and green. On the night of a lunar eclipse, I had a diffuse aurora that spun around me overhead 360°. It was stunning and yet quite colourless, though that didn't make it any less special.

The guesthouse in Grundarfjörður was comfortable and was a short walk to the harbour, where I had an early morning appointment. Laki, the tour boat owner/captain, had been helpful when we had spoken on the phone, and he assured me that there was room on his tour for me. He was late for our meeting and arrived looking as though he hadn't slept. He confided that he had a lot on his mind. We headed to the café on the harbour, which was rapidly filling up with tourists here for the whale-watching tour. This was the only tour in wintertime offering a chance to see orca or killer whales. I was thrilled to be taking part and excited at the promise of what lay ahead. The morning was freezing cold, around minus 8°, and our small group was thankful for the overalls provided to keep the elements out and us afloat in the case of an emergency at sea. The boat was smaller than I expected, given the rough seas, but I was confident we were all in safe hands. The safety briefing and introduction to the tour were first class and given by a qualified marine biologist who would accompany us the entire day. I gladly obliged when asked to carry the hot chocolate supplies to the boat. This area of Iceland attracted huge shoals of herring into the fjords during winter, making this a perfect feeding ground for whales and sea birds. I was hopeful of seeing orcas and had my camera ready – together with my woollen hat, two pairs of gloves, daypack with water and snacks, toilet paper, binoculars, spare camera batteries, hand warmer sachets, sunglasses and, believe it or not, sunscreen.

A small group of less than 30, we sailed out of Breiðafjörður Bay just after eleven o'clock in the morning. Daylight had appeared a short time ago and would last till around three o'clock. The fresh air made me constantly hungry, and despite my huge breakfast, I tucked into the small sweet pastries provided and a mug of coffee from my thermos. I saved the offer of hot chocolate for later.

Suddenly Laki and his crew stirred into life with shouts of, "Whale!" Memories flooded back of unlucky John and Mary whom I'd met when sailing to Iceland. Approaching us were numerous tail fins, all on the same trajectory. It was surreal. As one fin turned, they all turned and twisted, hunting in a pack, herding the fish below the surface into a ball, ready for their second hunting phase of feeding. This entire action was known as carousel feeding. The air was full of sea birds hoping to gorge on head and spine leftovers. Tears of joy sprang to my eyes, and I wasn't alone. We saw baby killer whales with their jet black and peachy orange colour (not yet turned white), a rare sight on the tour, according to our guide, while the females of the flock guarded them closely. As the tour progressed, we saw more and more orcas together with a small pod of white-beaked dolphins and a "flying door" (Haförn or sea eagle) with its colossal wingspan of well over two metres. As the light began to fade, it was time to head back to the harbour. I was frozen but elated. I took my warming cup of hot chocolate on the outer deck and enjoyed the backdrop of the majestic mount Kirkjufell in her winter coat.

As I drove back over the mountain pass towards Borgarnes, I felt sleepy from the heat in the car. I turned it down and pulled on my gloves and hat instead. I thought back to the light display the evening before, the magical day I had just had and felt suddenly overwhelmed by all that I was seeing and doing here with no one special to share it with. A waterfall of tears came from nowhere. I had been on the road for days and now craved the comfort of my small room with its shabby furniture once more. Blönduós would, however, have to wait another day as I had one more stop to make. The road stretched ahead, and I knew there was a huge number of sights I still hadn't visited. It was impossible to stop everywhere. Christmas wasn't so far away, and I decided that, if the opportunity arose, I would take another tour here as a treat for my eldest sister. Since losing our grandparents and parents, we had become a tight-knit family, and wherever we were in the world, we tried to spend Christmas together.

As I drove onwards in the dark, I suddenly was overcome by fatigue and stopped the car. I had no energy to continue or to even find a place to camp. I simply locked the car doors and lay down on the back seat with my sleeping bag. I slept the entire night. The following morning as I made a warm drink, for the first time in many years, I knew I had turned a corner. I was on the road to recovery. Yesterday's outpouring of tears and tiredness had a cleansing effect. I had gone to sleep feeling physically heavy and distressed and had woken up and instantly felt a change in myself. I had plenty of time before my appointment, so crawled back into my sleeping bag and knitted a square, choosing the brightest colours I had to hand.

As daylight left Húsafell, so did I in a heavy snowstorm, which had followed us off the glacier. Waving goodbye to the charming guides that had given me a fun and interesting tour inside the glacier, a marvellous piece of engineering which had become a huge tourist attraction, I promised them I would pass on their regards to the team at Blönduós search and rescue, several of whom they knew personally through their own volunteering. It was almost dark, and the pale new moon offered no extra light. As I passed the layby that I had parked in the previous evening, I noticed over the low wall the river below and an impressive fish ladder built to help salmon travel upstream. Its steps glistened silver in the low moonlight. Having stopped and positioned the car with headlights full on to take a picture, 50 metres ahead of me, a pair of snow ptarmigan crossed the road, stopping to peck at sand. I managed to take a quick photo of them just as they took flight.

It felt comforting to finally arrive home to the historic lady's college and let myself in by the wooden front door. I caught a strong whiff of the Kill House; some things don't change. It was ten o'clock on Saturday evening and the house was noisy and bustling with new faces. Along the corridor, there was a pungent smell of something boiling away on the stove, lentils with something sour. I decided to steer clear of the kitchen, brush my teeth in my room without rinsing and crawl into my single bed. I'd missed last orders at the pool as I had unpacked the Toyota and delivered it back to the garage. I looked forward to catching up with the local crowd in the hot pot on Monday (as they now closed on a Sunday). As I lay in the uncomfortable bed, I let my mind make sense of the past few days. Multi-coloured skies, whales, icebergs, gravel roads, and glaciers. The scarier moments driving on two wheels, the graffiti that lurked around unexpected corners, the tour in the front seat of a converted missile carrier over Langjökull over the glacier, and a patchwork of colourful knitted squares slowly making up a blanket of memories. I'd survived this far and hoped my luck would continue to hold.

ICESAR

The most basic of Monday afternoons could turn into something special without warning. A casual conversation took place, and wheels were set in motion. I had mentioned to a lady who I frequently met in the pool that I was thinking about publishing a book inspired by my trip, and if it turned a profit, I would like to donate some proceeds to the local Blanda branch of ICESAR. She was genuinely excited and knew someone who knew someone. That was how things happened in Iceland. I stressed that there was no guarantee of selling many books, in fact, there was probably more chance of seeing the lights in the summertime, but I still hoped there might be sufficient sales to warrant a worthwhile gesture. Following on from that brief chat, I received a phone call from an ICESAR representative in town. He invited me along to the Wednesday evening meeting to meet the crew and introduce my idea. They had also recently taken delivery of a new vehicle and were planning to run some trials in the highlands (interior) the following weekend. Depending on the volume of their call-outs and available seating, there was an invitation to accompany them. Worrying I might get in the way, he assured me they needed a "dead body" to carry on a stretcher. Hah!

ICESAR was the Icelandic Association for Search and Rescue, known locally as Slysavarnafelagid Landsbjörg. This coordinated Iceland's expansive web of emergency response volunteers who, despite having routine jobs and private lives, offered their services to keep Iceland safe. They attended all manner of incidents and disasters. During that first meeting I attended in Blönduós, they introduced each team member individually, listing their specialised skills. I learned that their combined competencies came in handy rather too often. ICESAR had around 10000 volunteer members, and at any one time, around 4000 people were on call-out duty over the entire country. This mass operation of vehicles, equipment, and training required funding. In this inhospitable landscape with volcanoes, earthquakes, crevasses, avalanches, tidal waves, glaciers, gale force winds, flooding and landslides, not to mention oceanic rescues, it was little wonder there was a need for skills like parachuting, nursing, mountaineering, glacial trekking, hunting, driving, and vehicle mechanics.

Unselfishly, they also helped overseas as part of a quick reaction force. One of the ICESAR units was the first response team to arrive in Haiti following the earthquake of 2010, landing in Port au Prince 24 hours after the mega quake struck. They took with them all the equipment and supplies they needed to allow them to work for an entire week unassisted, which included 10 tonnes of equipment, 3 tonnes of water, all telecoms, and their own water purification system. Icelandair sponsored them by supplying a plane from Keflavik International Airport.

They stressed to me the need for funding, and I thought of how similar it was to our RNLI back in the UK. Here in Iceland, most funds came from the sale of fireworks and private donations. Every year, ICESAR brought out a souvenir key ring called Nedyar Kall (Emergency Man), which people loved to collect. I was given one as a gift from that meeting, and since then, have started collecting them myself. My conscience wouldn't accept it without making a small donation, though.

ICESAR could trace its roots back to the Westman Islands in 1918, where the women banded together and organised a rescue crew to curtail the loss of their men at sea. This was soon followed by other communities who established similar protocols. I could relate to that, coming from a small island that had a history of lives lost at sea. It was difficult to talk about ICESAR without mentioning the effect tourism has had on their resources. Visitors accounted for more than three times the size of the population; their numbers exceeded well over a million a year, and it was fair to ask: was the service being used for its original intention, or was it an underfunded, underappreciated national concierge? Iceland marketed itself as a country full of raw nature and adventure, it attracted extreme sports enthusiasts, and it was little wonder tourists got into trouble. Speaking at length to my local team from Blönduós (Blanda), it quickly became evident they, too, were tired of tourists getting stuck in their vehicles through reckless and stupid behaviour. They were professional to the letter, however, and said, should anyone at any time require help, they were there to serve.

Having met them, I was immediately impressed by their commitment to the cause, and I looked forward to accompanying them out on their vehicle trials. Upon bidding farewell, I promised I would drive carefully and not intentionally put myself in harm's way. They all confirmed they had seen me around the area and had wondered what on earth I was doing there. They were glad to finally meet me.

The following morning, I was feeling housebound as I had stayed in my room writing since I had returned from my long road trip. I only ventured into the residency's kitchen and dirty bathroom when I absolutely had to. Most days, I showered at the pool and ate at the N1 petrol station when meat soup was on the menu. The artists and I shared dinner occasionally at the residency and tea with cake on knitting evenings if I was home. I decided to head out of town, about an hour's car drive, to my favourite beachcombing area, Hvítserkur, famous for its basalt sea stack, known under many names, including Troll Rock. The journey took a little over an hour along a series of small and bumpy unsurfaced roads, full of potholes and covered with a sheet of ice. It was a sub-zero day with a biting cold wind. I parked on the headland, pulled on my heavy leather mittens, and headed down to the beach along a well-traversed path, taking care where I placed my feet so I didn't slip. My foot injury hadn't healed as well as I'd hoped, and I still had pain and was unable to put all my weight on it. I wasn't a good patient, in fact, my family told me the only pain I accepted well was champagne ... and they were probably right. Despite the wintry conditions, it had felt good to be outdoors, and I slowly made my way along the rugged beach to the large stack of rock. To my far right, at the mouth of Sigríðarstaðavatn, the seals, which usually hung out on the rocks like giant grey bananas, weren't there. However, the recent weather had brought in all manner of flotsam and jetsam, together with a half-eaten carcass of an enormous ugly fish I couldn't identify. Eventually, numb with cold, I turned my back to the wind and started to make my way back the way I had come. I knew if I walked briskly, it would take a good 40 minutes. I had little energy despite eating well earlier in the day. The wet sand had made walking difficult, and my injured foot was throbbing painfully. As I walked on, I considered the steep cliffs to my right and wondered if there was a shortcut I could take to reach the car more quickly. The small waterfall that tumbled down the cliff was part frozen, and the rock face was dripping wet. For the love of God, I have no explanation as to why I decided to tackle the cliff, believing it wasn't so very steep and confident I could pick my way up and over it. I started climbing sure and steady, ensuring I had a good hold with each move. I had progressed a good 9 metres before realising I was stuck. There was simply no foothold or handhold with which to get a secure grip. My large gloves hindered me, and I had frantically shaken them off to remove them, one by one, daring not to let go with my other hand for dear life. My feet were sliding with every movement, and I knew it was impossible to try to crawl back down. To let go meant a certain fatal fall. I mentally lost control and panicked, shaking and sobbing, knowing there was a good chance I wasn't going to survive this. No one could help me now. I couldn't even reach into my jacket pocket for my phone as I knew I couldn't hold on much longer with my frozen fingers. I tried to work my way left, but it was impossible as there was a wall of solid ice. Somehow through my blind panic, I eased my hand into a razor-sharp slit in the rock to my right. I was terrified to remove my right foot but knew I had to, and as I did so, I drove my knee against the rocks as hard as I could, allowing my trousers to snag, giving some slight traction, eventually digging my foot into anywhere it could rest. Quite how, I don't know, but I did find a foothold, then another, and somehow inched my way sidewards and upwards to the top. There was a grassy verge above my head, and I reached up and dug my fingers in and dragged my body up onto terra firma. The nightmare was over. Shaking from both fear and cold, I lay there for some time, barely believing what had happened. I knew I was in a state of shock at how near to dying I had come. I stumbled across the headland and found my parked car. As I fumbled for my car keys with my hands trembling, I noticed blood and felt wet warmth on the side of my face. I sat in the car and nursed myself. I had hot water in my thermos and added three small packets of sugar I had collected from the café, added two Paracetamol tablets and drank it. It was nearing two o'clock and already dusk. I felt acid surge up into my mouth and opened the car door to vomit. I didn't have the strength to unpack my first aid box to clean my cut face; I would do that when I arrived home. I also made a mental note in future to place it where I could reach it in an emergency. I checked my handbag was beside me and fastened my seatbelt, then drove slowly back along the rough road to the familiar lights of Blönduós, managing to control my panicked state until I arrived back to my small room.

The following morning on my way to the post office, I met one of the organisers from ICESAR. He and his team had responded to a middle-of-the-night call-out and were now on their way home to shower, change and go to work. He noticed my cut face and asked whom I had picked a fight with? I explained where I had been and what had happened, and he was genuinely shocked I would try something so stupid. I could tell he wanted to say he thought I was smarter than that. Having listened to them tell of stupidity and recklessness that had put theirs and others lives in danger, I now stood in front of him explaining I had done the same thing by disrespecting nature and the elements and imagining I was capable of more than I was. In truth, I was so embarrassed by my actions that I avoided further discussions about accompanying them on training days – and more awkwardly, I wasn't invited.

HOT POTS

By early December, the daylight had decreased noticeably, and the shortest day of the year wasn't far off. Blönduós had grown up where the river Blanda cut inland from the sea. The rough landscape of rocky moors and tussock alpine and salt-marsh grasses offered little other than grazing land for horses. The town sat low in the bay and was shadowed by hills to the south. This meant that by early afternoon, the sun had already dipped out of sight. In fact, many days saw little more than a light halo in the distance, a silvery pink hue reminiscent of a David Shepherd painting. The sheep would have completed the illusion had they not been indoors for the winter. The days might have been short, but the nights were long, yet despite the bright moon hampering my efforts, I carried on searching the skies. The "lights" had been quiet, but in recent days, aurora activity had increased. I couldn't say the same for the town. The N1 petrol station was much quieter. It had been weeks since I had heard foreign accents and had started to miss the transient tourist trade. Stopping to fill up the Volvo with diesel, I met only locals and lorry drivers. Almost every car was a 4×4, and many I noticed didn't turn off their engines while they refuelled. I was seriously unsure about the risk of explosion, so asked if this was safe. They assured me it was, probably, followed by "Haa!". The staff mentioned they were more concerned about the number of people who drove off without paying for their fuel. So much so that more and more garages had been forced to install a pre-payment operation. I had been charmed in Reykjavik some months before by the petrol station with personal service, not unlike the garages we had in the UK during the 1970s before they all became self-service. Not having to get out of a warm car on a wintery day was worth paying a little extra for, and they even checked my screen wash and oil for free. The elderly gentleman who had attended me in Reykjavik might not have spoken the best English, but he enthusiastically recited "*niutiu og fimm*" in around 12 different languages for me. Eventually, I realised that he meant the fuel octane, and I became more interested in reminding him several times that my Volvo took diesel and not 95.

It wasn't only the garage that had lost trade. The campsite and guest houses looked closed, the residency had only a handful of artists staying for December and January, and even the swimming pool had lost its usual buzz though they assured me that after New Year, the local weight watchers club would be full of new and renewed memberships and the gym would become well used. In the meantime, I made the most of having the pool to myself and never tired of the pleasure of swimming in scrupulously clean fresh spring water, which was not over-chlorinated but just enough to maintain hygiene. A sophisticated underground system regulated the quantities depending on air temperature. There were no tidemarks of scum on the tiles that I'd seen in municipal pools back home. The entire place was spotless and affordable. They remained open the entire winter from early morning until around nine o'clock in the evening. The local pool in Blönduós was proud of regularly holding the title of the cleanest pool in Iceland, and justifiably so. All bathing facilities were outdoors, and the quick dash between the shower and the pool took a little getting used to. I gave up on flip-flops and preferred to run barefoot through the slush and ice, though I kept my wool hat on the entire time. I went through the same calculation every evening. While swimming a few lengths, I'd decide whether I could bear taking the metal steps out of the pool, which meant a longer walk to the hot tub, or whether to unceremoniously climb out and risk scratching my knees on the tiles to save the extra couple of metres walk. It depended greatly on how much snow had fallen and even more on how windy it was.

The run from the pool towards the two hot pots involved not only trying to look casual but also not falling over in front of everyone. While making my way to the hot pot, I needed to make the snap decision which hot pot to get into depending on how many people were already there. There were two to choose from. The hot tub furthest away was 37° and the nearer one a blissful 41° in temperature. Like most people, I wanted to sit in the "hot" pot, although there was a strict etiquette as to how busy a bath was before you just didn't join in. Once in, it was imperative not to touch toes or any other body parts of fellow bathers. I adored my evening ritual, which had the effect of warming and relaxing me to the point of sleepiness. Sadly, the last few minutes each evening were spent psyching myself up to run the cold gauntlet to the showers, then dress in heavy clothes and boots to walk home.

My favourite evenings were those that saw me submerged in warm water, looking up into the sky to see the magical palette of lights swirling overhead. On more than one occasion, I screamed out "lights" to no one in particular. The locals would glance upwards, shake their heads, and say "hah!" (the type of Hah! that translated as "it's only a 5 out of 9"). The kind staff in Blönduós, however, got used to me, and when an aurora started to appear, would turn off the main floodlights for my enjoyment. The locals, however, still thought I was mad.

It wasn't only the town pool apparently that could offer submersion under the starlight. In the N1 diner, I overheard three men talking about a grotto. That same evening in the hot pot, I asked the local group of policemen if they knew the place. Of course, they knew it well, they said. It was around a one-hour drive but not suitable for my Volvo, as the unpaved roads could be difficult. I don't know why I was surprised they knew which car I drove. Either way, I must have looked disappointed as they rapidly talked among themselves in Icelandic, then looked back towards me and in English replied, "Ok, we would like to help you". They nodded in unison, and I waited to hear more of their plan. As luck would have it, one man's wife was going in the general direction of the grotto at the weekend for a goose party. I could drive their car, a suitable 4×4, he added, drop his wife, Hulda, off at her party and then drive to the grotto. I would then collect her later in the evening and drive her home. He added quietly that he could go reindeer hunting with his friend as he now didn't have to drive her himself. I nodded in agreement and thanked them, wondering not for the first time what a goose party was.

Saturday evening came, and it was almost eight o'clock – not that that made much difference, as it had been dark for hours already. I dropped Hulda off at her party and headed in the direction she had told me to drive. After a short search, I came to a heavy gate, which I opened, drove through, and shut behind me. I wasn't sure why I shut it as Hulda's husband, the policeman, had assured me very few people came here in the wintertime, but it somehow felt safer. I parked, turned off the noisy engine and got naked, apart from my woollen hat, which I kept on. It looked like an ice rink outside, and even the short shrubs hung with florets of thick rime frost where the moisture had frozen. Having abandoned the warmth of the car and deciding against swimwear, I made an uncomfortable plod over to the cold hosepipe shower next to the pool, bit the bullet, took an ice-cold shower, then plunged into the promising warm hole in the ground. The water felt slimy, and my feet scraped the rocky bottom of the pit. The grotto was located beside a sea wall, with slapping waves and an outline of an island in the distance. Icelandic folklore said that Grettir was an outlaw who hid himself on the island of Drangey, a small uninhabited island with only a bird colony. A bit like Surtsey, I imagined. The story went that Grettir would come to this grotto to warm up, occasionally steal a sheep for food, then swim back to the island to evade the law. There was an honesty box nearby the pool to leave a token payment for the experience. Its upkeep required maintenance, and the simple pleasure that it offered was well worth a contribution. I felt brave to soak in a natural hot pool naked, though, as always, slightly conscious of being alone. My connection with nature always felt heightened by nakedness and darkness, and while I had taken a little time to get used to it, I enjoyed challenging myself. When I did come across out-of-the-way places and was alone, it felt like the perfect opportunity to be a little risqué. I'd closed the gate too, therefore was alarmed when stepping out of the slippery pool, I saw car headlights approaching. They were heading directly toward me with their lights on full beam. Fuck! I stumbled on the slippery rock as I attempted to race back to the car, and I felt like a hare caught in the headlights. Undignified, I picked myself up off the ground and rushed to retrieve my towel from where I had abandoned it carelessly on the way in, which by now had stuck to the frozen earth. The new arrivals, a middle-aged couple, waved and casually asked in a southern American drawl, "How's this thing work?" After a few minutes, I finally extracted myself and got into the car, not before explaining that it was supposed to bring them luck if they bathed naked as I had done. I hastily dried and dressed and, as I pulled out of the parking area, I was thrilled to glimpse two large white bottoms running over the ice towards the cold hosepipe. I might have damaged my elf karma telling a wee lie, but it cheered me up.

I drove back to Sauðárkrókur and the goose party to find Hulda a little worse for wear, which was also quite fun as she was usually so prim and proper. However, I was a little disappointed to find that the goose party was just a tame hen party with a much cooler name. When I mentioned I had unexpected company at the grotto, Hulda frowned and simply exhaled a low toned Haa! (the kind which meant "If you say so").

The past four months had converted me into a bathing addict, and a small interruption such as bad weather was not going to deter me from taking my daily dip wherever I could, though I did now prefer to keep my swimsuit on. Blönduós had been hit with heavy snowfall, and despite several attempts, it had become impossible to dig my car out, which resulted in me being town-bound. Temper tantrums seemed to escalate quickly among a few of the residents at the school, and I gave the common areas a wide berth. Most days, I took to roaming the outer limits of the town, following the river inland, then doubling back and approaching the town from the west over farm tracks. I gave up waiting for daylight or worrying about the dark. My route changed according to the weather conditions, and I still avoided the western approach, should the wind be easterly, to avoid the Kill House stench. I took my towel and swimsuit in my day pack, calling in at the pool on my way home. On the coldest days, the staff brought me coffee outside. They handed me a flimsy plastic cup and filled it from the heavy insulated thermos. I hung on to the cup with my fingertips while trying to keep as much of my hand, arm, and the rest of my body underwater, which was cooler than usual but still a pleasant 35°. On one day during a particularly cold snap, the milk blew away and froze before it hit the cup, the coffee blew to the side of its plastic crater, and I managed to aim it into my mouth and slurp greedily, narrowly avoiding contaminating the pristine pool. Just as I finished drinking, I relaxed my grip a little, and the cup became airborne, and I had to brave the elements running through the snow to retrieve it, place it indoors then again run back to the warm pool, hoping no-one had seen me. Hot drinks in the pool were a rare privilege as there was a strict policy of no eating or drinking in the bathing area. Sadly, having mentioned this to one of the ladies who was spending a month with us at the residency, she took her duty-free vodka and started drinking in the pool, inviting the shocked locals to join her. A new sign reinforcing the "no drinks" rule was set up, and no more coffee was offered.

During the snowy lock-down, there was little I could do other than knit and write. I cleaned the accommodation, reorganised my room, and caught up with my washing. On several evenings, I joined the ladies who gathered for prjonakaffi (knitting coffee) and was delighted when they commented that I held my knitting needles more naturally now. I still couldn't knit with a circular needle, but they assured me it was fine to use two straight ones instead. I searched the charity shop and supermarket for any discounted wool for my blanket squares, while at the same time stocking up on provisions in case the weather deteriorated further. Several times I had found it difficult to buy fresh fruit and vegetables, and though I still had several packets of crackers, oats, and tinned sardines together with coffee, cereals, and chocolate, I had eaten very little in the way of fresh produce, meat, or fish. The only fish I had eaten in Iceland was pickled, dried or in a restaurant. I started to concern myself over the possibility of both vitamin D and C deficiencies. How fast could scurvy set in, and how would I know if I had it? I broke open the multivitamin bottle I had brought with me for emergencies and chewed two tablets. *No point taking any chances*, I thought to myself. Food was expensive, and each time I had stood in the supermarket and calculated the price, I talked myself out of buying many items. Food, too, was still disappearing from the kitchen in the residency, and I had taken to storing expensive items in the car, though now everything in the car was frozen and I couldn't get the car door open anyway. The snow continued to fall, and I took it upon myself to shovel constantly, trying to keep the path to the residency clear. The snow fell so fast that within the hour, I couldn't see where I had been, and I was not sure why I even bothered. I'd pause to stretch my back from time to time and watch the small seals braving the icy waters in the small pockets of the river Blanda that hadn't yet frozen over. They reminded me of the two pet seals I had had as a child on our island home. I'd named them Plum and Custard. They would crawl into the tin bath we left outside and sleep. From the snowy path, I watched the small heads curiously pop up for a quick breath, then just as quickly dart below the icy waters to feed. I wondered how the seal colony I saw at Hvítserkur managed through the wintertime but guessed their thick blubber was a useful insulator. Watching them play during autumn, the young pups looked curious and playful as, with minimal effort, they scratched their skins free of barnacles and unwanted parasites on the shingle beneath, enjoying free rides on the crest of the waves.

Having grown up on a beach, I have always loved seals and I'd found that beachcombing on the beaches, especially on the western side of Iceland, was a perfect spot to enjoy both. I found it hard to walk without stopping every few steps to bend down and pick up something, turn it over in my hand and move on. The black sand beaches, and even one red sand beach on the west coast, were my favourite places to head to, especially after bad weather. Since my episode at Hvítserkur, I hadn't gone back there, choosing other beaches with better memories. Who knew what the tide could wash up, and though I never found Ambergris, I did settle for interesting shells, pebbles, sea glass, and driftwood.

CABIN FEVER

I often had to remind myself that I came to Iceland for something more than just a holiday. Most days, I took time exploring, photographing, writing, knitting or even sewing a little of the tapestry. Each month in the residency, there were arrivals and departures, and despite using the room only as a base, during bad weather, I was forced to stay put for longer than I had hoped. I found that the small group, myself included, obsessed over who would arrive next, how they would settle in, and whether they would fit into our group dynamics. Some people took up more space than others. I hoped they were tidy and prayed they would be quiet. Those visitors like myself who had never shared communal living seemed to take a few days to find their feet, while those more familiar with it came armed with lentils and beer and headed straight into the busy kitchen to make themselves at home. I was the longest-staying "artist" in residence, yet I was still the outsider. Strangely I felt much more at home in the company of local people, contrary to the other visitors to the school.

The textile residency attracted people from around the globe, and in the few months I was there, I met Canadians, Americans, British, Australians, Belgians, Dutch, and Germans. A few came accompanied by family or friends, though most arrived alone. Around two o'clock one afternoon, a noisy group of Australians arrived. They had been transferred from another residency not far away that had suddenly become overbooked and needed their rooms, so now we had to accommodate them. The two women and two men were dressed in the most preposterous tie-dyed attires I had ever seen. The largest woman, pacing ahead of the others, had bare legs and wore a flowing rainbow cardigan that was soaking wet where it had trailed the ground. Her long lank hair, combed into a traditional hippy parting, was dyed yellow and green – I thought she resembled a fast-moving aurora. As she wiped a finger over her enormous steamed-up glasses, she loudly voiced, "We're staying for one month", then forced her large bottom through the small doorway into the kitchen. As a Canadian woman and I finished off our lunch, the entire group squeezed into the cramped room, one dumping his carrier bag directly onto my side plate on the table. Before I could say a word, his filthy man-bag banged me on the head in his haste to look inside the food cupboards above me. The aurora brigade was difficult to live with, and hearing "Karma, man" from a spotty man with filthy dreadlocks made me grit my teeth. The group might have taken over the residency, but the local pool refused them entry for not showering. Elf karma ruled in Iceland.

Sharing living space with others was something I struggled with. I'd coped by staying out of the way as much as I could. The huge iron radiator in my room and the view made for comfortable living, and my bedroom door locked. There was access to a washing machine, which I needed, while the kitchen, functional at best, I used only if the house was quiet. I'd then cook a huge pan of soup or a casserole, which lasted me for days. The knitting group downstairs with their tea, cake, and conversation saved the residency from being a bad memory. Four months of tongue-biting and deep breathing not attributed to mindfulness or meditation was taking its toll on me. I often closed my door and thought, "What the hell am I doing here?". I might have progressed on my own journey of self-discovery, but no amount of healing or relaxation would make me enjoy communal living.

The first day of each new month brought with it new arrivals, and it felt like the beginning of a new school term. Generally, people arrived with a positive outlook and full of enthusiasm. I struck up an unlikely friendship with a few artists, though couldn't imagine staying in touch.

There were, of course, those that I hoped I should never meet again, and if I did, it would be a day too soon! A volley of arguments over dirty dishes and filthy toilets, food stealing, and monopolising the weaving looms were played down the long corridor. Small irritations became larger bugbears, and these disrupted the harmony and made for troubled living conditions. Being an artist wasn't a nine-to-five job. Most worked either when light was useful to them or to the tune of their own body clock. Many were night owls, staying in bed until noon, then after a lethargic start, worked late into the night.

The accommodation comprised a main studio at one end of a long corridor, the kitchen and bathroom at the other end, with the bedrooms in the middle. The end of the corridor to the west led through a fire door to the open studio with stairs leading downstairs to the administrative offices and main entrance. The loom room was one flight of stairs up. This meant that to access either the kitchen, bathroom, or work studio, you had no choice but to walk past each bedroom door. Many people wore heavy outdoor boots inside, while others allowed the heavy fire door to bang closed behind them. Many younger residents barely out of college giggled their way through the evenings and nights, and there was often one voice that took precedence. Enthusiasm for one's craft or a particular penchant for sheep was sweet and even charming until you had listened to it loudly on repeat for a month without pause.

The facilities provided were basic, and I was thankful I had taken my pots and pans, knives, and crockery, not knowing quite what to expect. I had also taken my own pillow, duvet, and bed linens together with a few towels. These were luxuries that having my car allowed for. I had initially left my items in the kitchen for others to enjoy using, though eventually, after rescuing the last few items that remained, I took to storing them in my room. I had bought some decent quality olive oil, spices, and seasonings too, but these too disappeared. I never did quite understand why certain individuals arrived with the presumption, "It's not worth buying things as I'm only here for a month". Freeloading was the term I used without apology, and the aurora brigade were the worst offenders or thieves, depending which way you wished to view it. I was shocked when on his second day, the Aussie dreadlock man marched into the kitchen, picked up my bottle of olive oil from the box with my name on it and started to use it to cook with. I asked him kindly to put it back. I had only a small amount left in the bottle, and if I were careful, it would last me the remainder of my stay. "Fuck you," he shouted, and turned on his green striped stocking feet and marched out. "Fucking freeloader," I shouted in return. The bad atmosphere hung stiff in the air, and we didn't speak at all beyond that point, and I was glad to see the back of him. Over time, several ladies commented there were fewer and fewer cups and plates available. I used my own, so hadn't noticed. They all agreed the "cardigan" woman was stockpiling them in her room. They were wary, so asked me to intervene. I took great delight in knocking on her door and shaming her into bringing out the pile of dirty dishes. Several girls had taken offence to her behaviour, and this was their opportunity to collectively tell her so. At times like this, I wondered just what I was doing there, acting like a mother refereeing petty behaviour. It was also bringing out the less attractive side of my own character.

The darkness, too, was now tiresome. I felt as though I had brain fog, and some days felt drowsy. I put it down to the menopause and continued taking the multivitamins, together with a high dose of vitamin D that I'd bought at the pharmacy, in the hopes it would help ward off the constant sleepy sickness I felt. It did, and within a few days, I felt much better.

From Thanksgiving through Advent and the run-up to Christmas, the nightly festive spirit in the residency escalated, and there was only one thing for me to do: hit the road. The car had emerged out of its snowy igloo, and I had managed to de-ice the locks. Bothy was packed together with warm clothes, cameras, several bottles of water, and my storm kitchen. On the front seat was my map and handbag, and in my pocket: my phone. I had no plan, and excitedly drove out of the small town to the T-Junction. Left or right, I wondered. I chose left towards Akureyri as there were still many places I hoped to visit. I hadn't driven more than three miles when the road conditions suddenly deteriorated. I inwardly groaned, turned the music up a little and pressed on. I had checked the weather forecast earlier that morning, and it mentioned icy patches. By now, I understood that patches of ice translated to ice for several patches of your route! Even the wind chill affected ice on the roads to the point where, for days, roads were treacherous, then suddenly as the wind calmed down, the sun shone for a few hours and the roads cleared. You could never be sure what conditions you would wake up to. Icy conditions weren't usually a problem for my Volvo. On the main roads, it coped fine, it was the unpaved, uneven roads that were challenging. It was still dark when I left Blönduós, but at least I would be driving into daylight, so I continued with no plan other than to escape the school.

Slowing down at the junction, allowing the bin lorry ahead to slowly turn right, I spontaneously turned left towards Skagaströnd and then a few miles further on took a right and headed towards Sauðárkrókur, the same road I had taken to Hulda's goose party. I chose the coastal route north on the Troll Peninsula. The road was reportedly open, which was a positive start, as several times when I had planned to drive this way, the conditions had beaten me. I would stop first at Sauðárkrókur as I had promised Ragna that I would replace her circular needle I had broken during my first knitting lesson, and I needed to buy knitting wool. Hulda had recommended a knitting shop which offered much more than the supermarket in Blönduós. I might also have time to pop into Gestastofa Sútarans, a local tannery I'd heard about and who were famous for tanning fish skin. I was curious, and a couple of artists had begged me to buy some offcuts for them should I pass this way. Hopefully before daylight disappeared completely, I would make it to Siglufjörður and find a place to stay or camp. The dusky sky looked promising, and a sunny December day looked on the cards, which would do much to lift my spirits. Since first reading Ragnar Jónasson's novel *Snowblind*, I had imagined what Siglufjörður must have looked like. He described the idyllically situated fishing town as accessible only via a small mountain tunnel. I had to see it for myself. What I hadn't expected was for the tunnel to be one-way!

The drive went well, and by late morning, I followed the sign for Hofsós with its spectacularly placed swimming pool. The pool itself was hardly spectacular; neither was the town. A brief drive through gave me no reason to stop, so I made a three-point turn at the end of the road and drove back onto the main road and onwards to Siglufjörður.

I had not been driving out of Hofsós for more than five minutes when I braked hard, checked for traffic around me, then made a three-point turn on the narrow road. There was no other traffic, and I hoped what I had seen would stay still. I opened the car window, then drove at a walking pace back to where a young gyrfalcon sat on a wooden gatepost, its creamy feathers puffed up, its eyes scanning the barren land. I carefully reached for my camera and managed to take three photographs before it took flight. Over the months, I had seen a few of these glorious raptors and was surprised by the diversity of their plumage. They were majestic, and it was easy to understand why only a king could hunt with a gyrfalcon during the Middle Ages. Turning the car once again, I continued north. The entire route was stunning to drive, and the clear weather enhanced its beauty. The juxtaposition of gleaming, icy ocean and verdant hillsides in such proximity were breathtaking. The snow covered most of the vegetation now but made it even more serene.

It could strike you as unreal, but again I reminded myself, this was Iceland.

As I drove onwards, clouds settled around the tops of the foothills to my right, and to my left, the ocean was a mirror. Something broke the surface, not a cetacean; this was smaller in size but still impressive. I realised I was holding my breath and told myself to concentrate on driving and, at the same time, breathe. A few sheep were still on the low grounds, their coats looking wet, heavy, and incumbent. I hoped these would be taken indoors soon, safe from the harsh elements and wondered why they hadn't already been gathered in.

The road twisted into severe bends forcing me to slow down. The icy patches were back, and being close to the coast, the icy cliffs had melted then frozen, resulting in a two-lane skating rink. A rider on horseback coming from the opposite direction waved, his buckskin horse comfortably settled in a fluid tölt gait. He was wearing a Lopi sweater with an equine motif woven into the pattern in the blue and black colours of a glacier. With only a light waistcoat over the top, he evidently didn't feel the cold. My car heater was a comfortable 24°, and my seat warmer on a constant low. I wore three layers of clothing, with a jacket beside me to put on before venturing outside. I was envious of the Icelandic ability to cope with the cold. I argued their genetics notation given that my own traceable roots went back several generations of Scottish fishermen and farmers, making me no less susceptible to the cold or damp.

Picking up signs for Siglufjörður, the traffic grew heavier. The snake of cars and lorries behind me forced me to increase my own speed as much as I dared. They tailgated my Volvo, and I was relieved to be able to indicate and pull into a layby, giving them the chance to overtake. I took a few deep breaths, enjoying the views. A small herd of colourful Icelandic horses grazed nearby, their hoofs working hard at unearthing any goodness under the frozen water. A BMW braked hard and parked behind me, and I watched a group of Asian tourists hop out with their cameras. I opened my car door, and the wind shocked me. It hadn't felt so strong while driving, but now, having come to this exposed summit, its full force hit me. I managed to carefully get out and close the door, zipping my keys into my pocket. I never left my car keys in the car for fear of the door somehow locking and trapping my handbag on the front seat. I turned to speak to the young man now making his way over to me, and as I did so, fell flat on my back on the ice. The wind helped propel me, and I grappled to stand up before the oncoming traffic whizzed by my head. It knocked my confidence when these unexpected events happened, and despite the embarrassment, it also made me think each time of the consequences. I'd hit my shoulder as I fell, but put on a brave face. The group of youngsters laughed off the moment and explained they had wanted to say hello to a local girl, and could they possibly have a photo with me, as they hadn't had much luck meeting any Icelandic girls? Bless! I chose not to burst their bubble and instead went along with taking selfies. They asked me to recommend a hot spring they could bathe in, and I pushed the truth, advising them the only spring worth visiting was my ancient ancestors' hole in the ground at Grettislaug. I wrote the details down and signed the note Emma Grettisdóttir.

Despite my sore shoulder from falling on the ice, I gritted my teeth and got back into the car, waving goodbye to the boys. Ascending the last few kilometres to Siglufjörður on Route 76, I suddenly noticed a sign for a tunnel, or was it a bridge? I wasn't sure. All it said was "Einbreið", or one lane. I slowed and reversed to check I had read the sign correctly and not missed any vital instructions. The circular blue sign had a white arrow pointing in my direction of travel and a red sign from the other direction. I took this to mean that I had priority. I drove into the entrance of the dark tunnel. There were no traffic lights as such, though a single light above shone amber. I'd no idea what that meant, and there was no way of seeing if there was oncoming traffic. I had driven across many one-lane bridges throughout the country but never a tunnel. A car was approaching slowly from behind, and I had no choice but to proceed through the dark one-way tunnel. I gripped the wheel tighter, as though it might help me, and drove faster, hoping to get the experience over with as quickly as possible, chastising myself for not letting the car behind me pass and go first.

Shy of a kilometre long, there were several passing places to my left marked with an M. Thankfully, the cars I did meet coming from the opposite direction used these to allow me to pass. I was only relieved I had not met any lorries as space was limited and was thankful to drive out into the light having survived my first one-way tunnel. On exiting, I checked, and the yellow sign with red and black arrows gave no more instructions than the other entrance had. This was Strákagöng Tunnel, 800 metres long and the second tunnel to be built in Iceland in 1967. The rough carving inside the mountain was neither pretty nor light, but it did allow the town of Siglufjörður (Siglo) to remain open throughout the year rather than to be historically cut off from civilisation for a good five months. Understandably this area inspired writers of grim Nordic Noir.

With daylight quickly giving way to dusk, Siglufjörður came into view. The steep descent was sheet ice, and my winter studded tyres fought for traction. With care, I drove towards the town, noticing the car behind, a 4x4 also patiently keeping its distance. I spotted the avalanche control gates high up the mountain to my right and thought of the dangers facing the inhabitants here, given the main residential area was located directly at the bottom of several potential avalanche paths. This was just one of many natural disasters Icelanders averted on a regular basis, and no doubt this town, too, had an active ICESAR volunteer group. As I parked in the quaint town centre, in front of a charming old two-storey building undergoing renovation, I was surprised to see practical timber being used here. Most buildings wore the usual down-beaten look scarred by the elements, unsurprising considering Siglo, as it was locally known, was only 28 miles from the Arctic Circle and close enough for wintertime to envelop the town in darkness for several weeks. There was a noticeable difference here, even compared to Blönduós, where on a sunny day like it had been, we would have had a good deal more daylight. The long darkness must have affected inhabitants mentally and physically, and I was immediately grateful I hadn't settled here for winter. The snow came early, too, and by the time April came, I imagined you could be fed up with it. The tunnel remained open, but the roads were exposed and often closed. Even the practicalities of shopping or going to work could be hazardous. As daytime temperatures rose a little, nighttime came and dipped the mercury, a seesaw of freeze-thaw making the pavements and roads treacherous, as they were now. Salt wasn't used much on the roads. It would be an expensive luxury and probably useless against the wind, which blew drifts of snow across streets and roads rendering them quickly impassable. I parked and walked a little to get some fresh air and see the town, though after a short 20 minutes, was back in the car, shivering.

Apart from the wooden scaffolding, I also noticed timber telegraph poles. I hadn't seen any forests as such and deduced they were likely imported from their Scandinavian neighbours. Siglo was exactly as Ragnar Jónasson had described it. Looking around in the dim afternoon light with the town settling itself in for another cold evening in December, the road likely becoming impassable again soon, it wasn't hard to imagine murderous goings-on. I understood where this country's enviable literary heritage came from. Iceland hosted clubs for everything, and even in the smallest of towns, you found music clubs, chess clubs, sports and fitness, online learning and education, flower arranging, arts, crafts, knitting and, of course, book clubs. Iceland might be the navel of the world, but there was no excuse for boredom.

Maybe it was the weather that nourished the need to knit, and here was no exception. One extraordinary local knitting project was that of a 17 km long scarf to be threaded through the new tunnels joining the towns of Ólafsfjörður to Siglufjörður as a warm gesture between the two towns. Frída Björk Gylfadóttir, along with another 500 or so knitters, joined together to complete the project. Men and women of all ages knitted tirelessly, the youngest only 10 years old, the eldest 94. The scarf was completed, and the twin tunnels opened with the scarf symbolising their connection. The actual scarf went on to be dissected into smaller pieces and sold to raise money for local charities. Although it appeared to be only one tunnel when you drove through, there were, in fact, two: one northbound and one south. Northbound was a whopping 7100 metres long while the south was only 3900 metres. Relieved it was safer than the last tunnel I had driven. I looked forward to driving through it on the way to Akureyri.

I left Siglo and headed towards Dalvik via Ólafsfjörður. I knew each journey took longer than the miles and was fine with that. I started to think where I might sleep as I hadn't yet seen any accommodation open. Feeling quite unsociable after the recent days in the residency, I favoured using Bothy but had no idea where I could camp. I was in no hurry to return to Blönduós, and using my tent allowed me to stay on the road a little longer. My road trips were bleeding money, and if I could save on accommodation, that helped hugely. I enjoyed the feeling of independence that camping gave me, and I hadn't felt the cold too much. I pressed on and at the same time scanned for possible places to camp. I drove through the tunnel south, a much brighter and more pleasant driving experience than Strákagöng earlier in the day. The third tunnel I drove through, Múlagöng, was also dark, dingy and one-way, though thankfully, I passed only one car during the entire 3400 metres. They had the right to pull over but chose not to, then realising their error, stopped and reversed. I couldn't say I felt the love for Icelandic Einbreið!

I approached Dalvik and made out the small harbour where the sign advised me the ferry left for Grimsey, which was situated around 40 km off the coast. A recent fire had razed the island's church to the ground, and alas, no historic valuables or documents could be saved in the 20 minutes it had taken to destroy everything. Iceland immediately pledged to invest between 80 and 100 million kronor to rebuild the church using timber and even driftwood. Apparently, the chief fire officer's sister-in-law on the island lived next to the church and raised the alarm. An employee of the tiny local airport, too, rushed out with the airport fire engine, though by then, there was nothing left to save. Other than the church, there wasn't much to see on Grimsey in winter, so I decided not to visit at this time. Further along the street, there was a grocery store, a sign offering sea tours by rib, and a garage that offered MOTs. There was a café that looked cosy, but it was closed. I refuelled at the small self-service petrol station, having all but given up on finding a place to camp, when the lady inside suggested I take the next turn right. There was a campsite, and she was sure the shower and toilets would be open. She offered to ring them to let them know I would be staying overnight. I thanked her and drove on past the swimming pool and what looked like a school. It was hard to tell in the dark. The entire area looked closed, but I took a chance and parked the car, walking the last metres with my packs. Bothy was set up with ease, and each time I used her, I thought of Mats and grew fonder of her. I hoped she would be a companion on my travels for many years to come. I left my belongings where they were and went to investigate the service block. The door was unlocked, but the water was either frozen or turned off. I didn't care. I headed back to the car for my spare thermos flask, water, and gas burner and boiled some water, popping a chicken stock cube, a good handful of instant noodles, and two hot dogs in to warm up for dinner. I unpacked some dried fish, butter, cheese, and Knäckerbröd I had bought in Ikea, together with a jar of homemade cloudberry chutney from Ragna. I shuffled into my sleeping bag, fully clothed less my boots, propped comfortably up against my pillow and backpack and enjoyed dinner. For some reason, I slept less well knowing I had neighbours around me and wanted to be on my way before they woke up. Knowing Iceland, the entire town already knew I was there, but still, it felt a little intrusive camping beside the school in the middle of town. Given that I had received no services and no one came to see me, I didn't pay.

As I quietly drove away the following morning, I took a final glance backwards as I couldn't work out whether I'd camped in the campsite or on the school sports field. Either way, it had been quiet and comfortable, but in future, I would stick to non-urban campsites. It was too early for daylight while driving towards Akureyri. It wouldn't be light until around eleven o'clock. I was sad to be missing the views, but hopefully, it wouldn't be the last time I drove this way. I couldn't help thinking of the first occasion when I'd driven to Akureyri from the west: the chance meeting with Ragna's brother at the garage and, of course, seeing my first humpback whales. It all felt so long ago, and I no longer felt like a tourist here. I thought of the nights I had camped out, the challenges I had faced, the miles I had covered, the injuries I had sustained, and the close calls I had experienced. I looked forward to visiting the town again with its small, welcoming pedestrian zone with shops and cafés. I also needed to visit the camera shop to replace a UV filter that had fallen apart.

The wind seemed to have picked up as I approached the coastal area above Akureyri, and plastic debris covered the road. I hadn't checked the weather that morning as I'd been in a hurry to leave the small town before it woke up around me. I parked the car and entered the camera shop situated opposite the large public car park, where several people were already talking about a storm that was heading our way. The storm itself was not surprising, though hearing locals talking about it was. I had rarely seen them react to a weather forecast with such enthusiasm. I interrupted to ask for details and whether they had heard how the weather would be later in the day as I would be driving back towards Blönduós. In unison, they turned to me and replied "Haa!" (The type that means "Are you joking?"). They cautioned me not to drive under any circumstances. *This must be serious*, I thought. A further discussion ensued, ending with a promise from me that I would not drive onwards today. So having paid for my new filter, I left the store wondering what my accommodation options might be. I didn't wish to risk using Bothy in a violent storm. The Christmas holidays were fast approaching, and there was sure to be an increase in demand for the few hotel and guest-house rooms that were open. I imagined Icelandic hotels wouldn't miss an opportunity to raise the prices too. I walked through the town centre, enquiring in the guesthouses and hostels whether they had a room available. Those that were open were fully booked. Flights out of Akureyri had been cancelled due to the weather, restricting landings at Reykjavik in the south, and people were literally stranded. I made my way to the largest hotel in the centre where, for a hefty sum, I could have the last room. I paid the bill in advance and thanked them for the key. Once in my room and after a struggle with the hotel Wi-Fi, I finally managed to connect and get an updated forecast. Violent hurricane-force winds were expected tonight with constant wind speeds expected of around 37 m/s. The word "constant" was a worry. That meant gusts would be far in excess, and I knew from experience the road between Akureyri to Blönduós was no place to be driving in extreme weather. It was too early to stay in the hotel room, so I took my bag with swimwear and camera and headed out for a walk. There was sure to be a pool open where I could rely on hearing the latest gossip and weather news in the local hot pot.

Akureyri was a compact town, quite hilly, with quaint buildings and shops. A few houses had tried with their front gardens, something you didn't often see in Icelandic towns. The traffic lights, when turning red, weren't circular, as with the rest of the world, but instead, lit up a bright red love heart. I never knew if they were intending to spread a little love or wished to remind us to take our heart health seriously and walk more. Both worked for me, and I strode on, feeling the love and working up a healthy appetite. The smell of fish and chips catapulted me back home, as it always did. I looked around and found the steaming windows of a chip shop and dived inside. The elderly gentleman greeted me warmly and offered me a menu. He asked where I was from, and when I replied, snatched the menu back and said I must try the Haddock. I didn't argue with him and was glad when he served me an enormous golden crispy fish with hand-cut chips, peas, freshly baked bread, and locally churned butter, together with a huge mug of steaming hot tea. Coming back a few minutes later to check I had everything I needed, he went on to explain which trawler had landed the fish, the actual fishing zone and even the captain's name on the boat. It was every bit as good as the fish from my childhood. Although my family weren't fishermen, the local boys working the trawlers left buckets of fish at our door in exchange for the loan of our car for a few hours, mainly to court girls without prying eyes. The fish and chips in Akureyri would remain one of the top two meals I ate in Iceland.

I walked off my early dinner, taking the directions the man serving me had offered and found the pool at the top of a steep hill. A handwritten note on the door said, "Closed due to bad weather, try again tomorrow". Disappointed, I walked back to the hotel, wondering whether Blönduós would also have closed their pool. I doubted it. As darkness fell, the sky was ominous, and the wind had picked up. As I made my way through the hotel doors, the receptionist suggested I ring Blönduós in case they were worried about me. I hadn't thought of that but did so anyway, and Jóhanna was extremely grateful for the call as she hadn't known where I was and had wondered whether I knew a storm was coming. Haa!

Overnight the weather in the south had worsened, and by the time I had finished breakfast, the state of alert that had been advised for the entire southern region the day before had now been raised to hazard level. The National Police Commissioner himself had followed suit and raised the security alert in several coastal areas to hazard level. Hazard level meant that added emergency and security services were deployed with certain areas off limits where there was a danger to life. I hoped the storm wouldn't reach this far north, but every news channel suggested that it would. The worst of the weather was still some hours away, and the northern roads were open. I toyed with what to do. If I stayed here, I might well be stuck in Akureyri for some days, so, upon finishing breakfast, I decided to drive back to Blönduós as quickly as possible. I packed the car, checked out of the hotel, and drove over the mountain roads non-stop. The wind was forceful but no worse than I had driven in many times before. I arrived safely and sighed in relief as I parked once again beside the shabby garages with their corrugated roofs and let myself into the school. By the time I switched on my computer, all news channels were covering the drama in the south. The storm was doing some serious damage.

The storm first hit the Westman Islands, off the south coast, where I had visited some weeks before. It was reported that the low buildings had taken the full brunt of the storm, which was now referred to as a double hurricane. It took only a short while for the emergency services to respond to the dangerous situation and close the entire ring road, starting with Hvolsvöllur in the south to Reyðarfjörður in the east and continuing in every direction thereafter. The storm played out as forecasted, gaining momentum as it travelled over land. As the afternoon passed, the news reports became more frenzied; shops, pools and schools all closed early, and the entire country was put on lockdown. Several of us staying in the residency worried how much the old building could withstand and packed a grab bag in case we were forced to evacuate. I fully charged my phone, torch, batteries, and computer in case of power outages that were already predicted. I filled the car with diesel in case we needed to get away (though with the roads closed, it was more a gesture to make myself feel prepared than being of use). I talked two other ladies through a plan of action should any of our windows blow in, a real hazard in such an old property. No one working at the school contacted us, and the general feeling was of facing any crisis alone. I guessed they assumed we would be fine. I checked the entire building for open windows and doors and ensured everything was secured, just as I would do at home. By late afternoon, the online news reported that such was the increasing magnitude of the storm, such conditions occurred only every 20 years or so. Every rescue team, together with the civil protection authority, were on their highest alert throughout the country. By now, the winds had reached a staggering 90 mph with gusts even greater. To put the severity into context, Hurricane Sandy, which hit New Jersey, had winds of 80 mph. Reports were coming through of roofs and balconies having blown off buildings, boats had lost their moorings, and bus shelters in Reykjavik had been blown from their foundations. Flying debris was hazardous. Thankfully the public had heeded the warnings, there wasn't a person out on the streets, and, apparently even in Reykjavik, there was only one place open to buy food, and that was the hot dog seller Bæjarin, famous for his best lamb *pylsur* hot dog, who had vowed to keep his small kiosk open whatever the weather.

The residency fell unusually quiet. The two other ladies and I sat together in the large studio room as Blönduós prepared itself for a late evening assault. It was already dark, and the flagpole outside fought to stay upright – the flag itself gone earlier in the day. The old school building creaked as an odd aggressive gust caught it with an easterly blow. Significant snowfall was expected to accompany the wind, though so far it hadn't appeared. Those residents not already obsessed with the weather up to that point certainly were now. Each news headline and updated forecast was repeated along the corridors, and one girl offered sleeping pills to the rest of us. I declined, preferring to stay awake and alert. I had already seen some violent storms and had experienced more than I cared to on the south coast when my car wheels had left the ground. I guessed we all handled fear in different ways. The news went on to caution us that the weather we had previously experienced here was nothing compared to what was coming our way. My thoughts went to the search and rescue individuals who put their own lives at risk for others. I also ensured I had their app safely on my phone and ready to use if necessary. It was a comfort just knowing they were close to hand should we need them, and again thought how I would like to help with a donation if possible.

On the news, ICESAR's own project manager Jónas Guðmundsson compared the situation to 1991, when Iceland saw such severe weather that winds literally threw whole cars around and even took the roof off the national hospital. We had prepared as best we could and agreed we would all stay indoors with our grab bags and phones nearby. We busied ourselves, waiting for what might come. I boiled water and filled the thermoses so everyone could enjoy a hot drink should the situation here worsen. By ten o'clock in the evening, we were all quite frustrated and finding it difficult to concentrate on anything other than the news. It was severely windy, and the windows of the building rattled. I personally didn't think it was so very much worse than we had had several times over the past few weeks. We still had electricity, despite the odd flicker for a moment or two. By midnight, I decided to get some sleep and hoped I wouldn't be disturbed.

By the time I woke the following morning and looked out of the window, the weather looked a little calmer. I turned on the computer and checked the national news. Devastation reigned through the entire country, though in truth, for the severity of the winds, I had expected more damage. The south coast had indeed taken the brunt, but even so, surprisingly, the reports seemed to be of minor damage. The wind had stabilised and despite it being difficult to walk upright and having a lot of grit and dust blowing into your eyes, it was safe to walk outside once again. I was glad I had made the decision to drive west from Akureyri the morning before, otherwise I could easily have been holed up there for a day or two in a very expensive hotel room. The residency might not have been much, but for now, it was home. When I was there in my small room with the chit-chat of voices nearby and my door locked, I felt safe. This storm had again shown me that roads became impassable for a variety of reasons. Not only was it snow or drifting that forced them to close, but the wind was often the greater risk to life. In the past, when I approached a road closed to traffic, I questioned why there was little or no snow to be seen. This journey had so far taught me much about prevention and survival.

The Strætó Bus service, the main bus service in Iceland, was still not running and it would be another day before it started again. There was no point given the road closures, and safety was their highest priority. Given that the entire country had been hit with winds twice as strong as the criterion for hurricane-force, I was relieved they took responsibility for passengers and avoided travel. After the event, the weather forecasters deliberated over what had occurred, and a consensus was reached. According to the local news, after the event, we had not, in fact, suffered a double hurricane or even a single hurricane. Despite the wind speeds suggesting otherwise, because we had not had the violent thunderstorms or heavy rain which accompanied it, our storm was downgraded to a simple "polar low". I wondered again whether it was this gentle reporting of the weather that lulled many people into a false sense of security. Despite all domestic flights being grounded and the international airport being closed for some hours, services resumed quickly, and life returned to normal. Although the roads re-opened, I decided to stay put for a day or so before making my way south to meet family for Christmas. I thought about the very sparse research I had done before my journey began and my decision to bring my car. In hindsight, I would have traded in my car for a 4×4, but I would still have travelled in wintertime. As for Blönduós, I could have chosen a prettier town, but by now, it had grown on me and had earned a place in my heart. The residency had not done so, and I doubted I would foray into communal living with artists again. By now, I realised I didn't care too much about Surtsey or my upbringing either. It had made me who I was today, though I'd earned what I had and where I was by myself with only myself to thank. Iceland had charmed my mother, and Surtsey was simply a news story – albeit an impressive one. All I wanted at this stage was to survive the rest of winter here and decide the next chapter of my life, which included finding a new home.

Iceland was full of perils: unpredictable weather, unpaved roads, slippery lava and basalt, glaciers, the odd polar bear on occasion, strong sea currents, the wind, not to mention hospitals and police going on strike for better working conditions. ICESAR had told me many tales of grown men dying of hypothermia short distances from their vehicles or becoming separated from companions and barely surviving overnight, exposed to the elements, saved only by the rescue services finding them with not a moment to lose. The dark winters could fray your nerves. During our knitting coffee one evening, we were told the story of intelligent adult women who couldn't live harmoniously in the residency, despite there being only two people living here for a month one winter. Both resorted to sleeping with knives under their pillows, believing the other was trying to kill them. It sounded like a classic case of cabin fever to me. For those reasons and many more, Iceland, and more particularly Blönduós, was not on my list for house hunting.

I decided to head to the hot pot, my last trip before Christmas. By now, I was familiar with the local people's routines. I could accurately predict who would be in the pool at which time and on which day of the week. When in Blönduós, I visited the pool most days, though I preferred going later in the evening as once I had showered, roughly dried my hair and walked back to the residency, it felt like bedtime. During the shortest weeks, when daylight hours were minimal, it grew dark by 3.30 in the afternoon. My body clock fought the onset of Polar Night Syndrome and tricked me into thinking it was time for bed when, in fact, it was barely seven o'clock in the evening. Despite being tired, if I did go to bed too early, I'd disturb the other residents by getting up soon after they had retired. Such a contrast to my years of insomnia. I wasn't sure that Iceland qualified as suffering from actual PNS (Polar Night Syndrome) – other than perhaps the small northern towns I had recently visited that were situated on the cusp of the Arctic Circle – but my growing symptoms felt real enough.

Strictly speaking, the polar night was when the sun didn't show its face over the horizon at all, forsaking daylight for more than 24 hours. In Blönduós, there was a brief glimpse of light in the sky for a few short hours, though due to the town being shadowed by hills, the sun itself wasn't visible for most of the time. I continued vigilantly to take my vitamin D supplements. The melatonin deficiency was more difficult to treat. I had spoken to a pharmacist in Akureyri who explained that melatonin was a hormone produced by the body, and its release depended on the alteration between night and day. Any deficiency in such a hormone could lead to a state of irritation, nervousness, anxiety, poor sleeping patterns, depression and led to premature ageing and the onset of some cancers. Iceland apparently had a problem with this in wintertime, and statistics showed a high dependency on anti-depressants. I hadn't known of this before, and it gave me something else to obsess about other than the weather. I became so concerned that no matter the temperature, as soon as there was a glimmer of decent daylight, I headed out to the hot pot, as this allowed me to expose as much of my body as possible to the light of day. The pool and hot tub area were shielded from street view by a high wall, which again prevented the sunlight from reaching my body properly. This resulted in me having to get out of the warm pool, walk to the corner of the wall and hold my arms above my head in the hopes of attracting a few essential rays of light to my wrists and forearms for a minimum of 20 minutes daily, as recommended by the pharmacist. It took the staff some days to finally stop giggling and ignore me, though by now, they officially declared me stark raving mad. As far as I was concerned, they could call me what they liked; I wasn't taking any chances with my melatonin. When I did strike up conversation with other women in the pool, I asked them the same question, did they take supplements against the lack of light? No, they all assured me, nothing. I didn't ask the men as they generally kept to themselves and didn't talk to a single woman in the hot pot who waved her arms over a high fence towards the school.

In addition to the swimming pool and hot tub, there was a steam room, which I had enjoyed when I first arrived in Blönduós. Sadly, after the first couple of days, this became out of order and stayed that way for several weeks and I missed it. I had tried a DIY version of Kneipp Therapy, which was to soak in the hot tub, then take a quick plunge in an inflatable water barrel filled with ice-cold water. I wasn't sad to see that removed when it finally solidified during winter. By that time, it was so cold I only had to stand up in the hot pot for a few moments with the same results. In an instant, your nostrils froze, your wet hair formed clumps of ice, and your goose bumps erupted, turning you into a shivering mass in no time. When I had company in the hot pot, I merely nodded a polite hello, sat back, and allowed the Icelandic chit-chat to wash over me. I found it relaxing living outside a language, unable to grasp what was being said, thereby less troubled to think about it. I rarely interrupted a conversation – unless, of course, I saw the northern lights. I found it delightful when locals initiated conversation with me. I felt more and more a part of the community, in my own way.

The funky smell from the Kill House had abated a little. The other odorous heavy industry in Blönduós was the sheep's wool washing plant. It was with trepidation that I entered the factory and offices of Istex, which has become a worldwide name. I had been invited for a tour of the premises to witness for myself how the raw wool was handled. Delivered fresh from the farms and leaving beautifully cleaned and ready for spinning, the entire process was carried out in the small factory situated in an industrial area close to the N1 garage. The smell hit me before I reached for the door handle. Istex, or Íslenskur Textiliðnaður (translation: Icelandic Textile Industry), own the only wool-washing facility in the entire country. Every farmer throughout Iceland sent their fleeces here for washing. Istex sold their wool through international distributors, carrying with it the quality promise of original Icelandic yarn. They bought wool direct from the farmers and then processed it into yarn for knitting, weaving, and carpet-making. Lopi (the wool which gave the famous Icelandic jersey its name) was probably the best known, together with the equally famous Alafosslopi, Léttlopi, and other Lopi yarns for hand-knitting such as Plötulopi and Bulkylopi. Istex also have a spinning mill located in the town of Mosfellsbær nearby Reykjavik, but the entire scouring operation was handled in the small northern town of Blönduós.

Istex gave me an unrestricted tour of their premises. The wool was unique in composition as Icelandic sheep's wool consisted of two types of fibres. Firstly, a fine soft inner layer providing insulation and, secondly, a long, glossy outer fibre which was a natural water repellent. The two layers combined to create a wool famous for being lightweight, water-repellent, and breathable. Istex wasn't a big employer but was invaluable in keeping the heritage of Icelandic wool alive. Around 50 people were employed in total, and they were responsible for processing up to 80 per cent of the wool produced in Iceland. Istex in Blönduós sorted and scoured the wool, then sent it on to Mosfellsbær for spinning. Once there, the wool went through a range of processes to ensure it was thoroughly cleaned and was then quality graded. The machinery appeared, for the most part, to come from Rochdale in England, stamped "Petrie and McNaught 1963". This apparently arrived in Iceland soon after being built and operated in various sites before settling in Blönduós. The factory ran through the wintertime, as the sheep were sheared in early November. The machines ran 20 hours a day, then were turned off for four hours for maintenance and cleaning. The foreman, whose name I really should have remembered, explained the processes in detail down to the environmental factors of dirty water disposal. The smell followed me, though due to the good ventilation wasn't nearly as unpleasant inside the factory as it had been from outside. Tending these antiquated machines was a huge responsibility, and although most parts could be fixed with time and patience, the foreman did say that there were two essential working parts he prayed would function a while longer. They were irreplaceable. There was a plan in place to replace the entire machinery, although a date had not been set. I hoped for his sake luck held out. The foreman described his role as fun and challenging; it was obvious that he had found his dream job. As we stood admiring the industrial marvel, he suddenly inched closer, scratched his chin and lowered his voice. He had another secret to share. Without fail, every year, when the machines were turned on after their summer holiday, they took exactly five weeks to work properly. For no reason, they stopped working, then started, had hiccup after hiccup and generally were not in the mood to work. Then suddenly, overnight, they ticked over like a dream. It was the same routine each year. I asked, did he believe in ghosts? Absolutely not, he shook his head. Elf karma? I ventured. Perhaps, he answered as his nod said yes.

The heated air from the vehicle-sized tumble driers warmed me before I ventured back outside to the cold. Sheep had never particularly interested me, and the finer points of knitting yarn had evaded me, but now I understood a little more why textile artists flocked to this area. They might travel miles in search of their materials and inspiration for their chosen craft, but I felt sure that, should they arrive here, they would be justly rewarded. Embarrassingly, when at the end of the very lengthy tour I was asked did I have a question, all I could think to say was, "Do you know the uncle of the man who returned my handbag to me last September?" No, he didn't, and I had to say, he looked rather disappointed as he left me to exit the premises alone. I walked the longer route home in the hopes the fresh air would rid my clothes of the pungent smell of sheep. My newfound respect for wool saw me lying in bed later that night with my headlamp on, knitting. Other than my tent squares, I hadn't knitted much, though was determined to complete a shawl for my sister's Christmas present.

I had picked up some wool from a lady who worked tirelessly, dyeing her yarns with plant dyes from nature. The plants she used were both indigenous and imported and included moss and lichen, lupin, rhubarb, onions, and even cactus. The array of colours was resplendent and the quality of her wool remarkable. I chose a soft Merino yarn which was less scratchy than the traditional Lopi. It was fine and light against the skin, in a dusky yellow turmeric colour together with a stronger silver green made from birch tree bark. The light in my room was inadequate to knit by after three o'clock, and despite having bought an extra lamp from Ikea, I still strained my eyes. I, therefore, resorted to wearing a headlamp to knit or read by, and as I lay in bed craning my neck to read my knitting pattern, tongue out in concentration and a bright lamp on my head, I wondered what my family would think if they could see me now. My body clock, now completely out of sync from self-diagnosed Polar Night Syndrome, meant that I replaced sleeping with knitting. Counting row after row was therapeutic, and by the time the rechargeable battery on my headlamp gave up around four in the morning, I realised I was knitting properly – and loving it.

TWO'S COMPANY

The 13 nights leading up to Christmas saw the descendants of Trolls come out and leave either rewards or punishments in the shoes of children. The gift you received, good or bad, depended on how well you had behaved the previous year. Over time, the "Yule Lads" had evolved from murderous monsters that ate children into mere pranksters. Their main purpose was to scare badly-behaved children into being well-behaved before Christmas. Their pet was the Yule Cat: no normal cuddly feline but a wicked beast who happily ate children who hadn't received new clothes before Christmas Eve.

On my first day driving between Seyðisfjörður and Blönduós, I had stopped at the stunning lava fields of Dimmuborgir (Dark Castle) in Mývatn. This was home to the Yule Lads, where they lived in a cave. On the 12th December each year, they surfaced from their mountain lair one by one to conduct their deeds. Children throughout Iceland left their shoe on a windowsill before going to bed at night, and in the morning when they woke, checked whether they had received anything inside the shoe. A small gift was a reward for good behaviour, whilst a rotten potato warned them to improve. The Yule Lads then left and headed back to their cave in the order they had come out, while the last one, "Kertasnikir" or "Kurt the Candle Stealer", left on the last official day of Christmas, 6th January.

When we were young, our mother read us the book *Christmas is Coming*, written by the Icelandic author Jóhannes úr Kötlum. Unfortunately, I was constantly ridiculed as I shared my birthday with "Pot Licker" who loved nothing more than scraping out pots for leftover food. If the cap fits ...

I suspected we had a few Yule Lads lurking in the residency as milk regularly vanished from the fridge, and pots and pans disappeared into rooms. Was it only the Yule lads who were partial to a bit of Skyr, or did this include the artists, I wondered?

Christmas in the far north came and went. I headed to the "home of the latte-drinking wool scarves", Reykjavik, to meet my eldest sister, who was coming for a short visit. After taking the ladies from the residency to the N1 garage for a pre-Christmas lunch, I was waved off with season's greetings and a shopping list. I would be back before Hogmanay, as I had an invitation to join the locals for an evening of a fillet of foal and fireworks.

After the basic accommodation I had endured over the past months, it was luxury to check into the Hilton Nordica Hotel, and I had no interest in stepping outside its comfortable doors anytime soon. With good light and a comfortable bed, I might even finish the shawl ready for Christmas day. My sister Caitlin on the other hand had other plans. The alarm clock set daily for an early wake-up, she sprang out of bed, opened our door, and checked whether we'd had a gift from the Yule Lads. There was no windowsill on which to place our shoes, and we were hesitant to leave our shoes outside our door in case they were stolen. She delighted in opening the small bag hanging from the doorknob to see whether it was a potato or a present, a delightful touch, courtesy of the Hilton. By the time I'd dragged myself from the huge comfy bed, Caitlin was already dressed and heading downstairs for breakfast. I grudgingly followed, packed the car with what we might need for the day, and we drove out of the city before it was light.

During my time in Iceland, I had emailed family and friends, regularly updating them as to where I had visited and what I had seen. Understandably, therefore, when welcoming visitors, they had a long wish list of what they'd like to see. Caitlin's planned trip to Iceland in autumn had been cancelled, thus narrowly escaping a stay in the Hotel Blönduós! A keen diver, she was desperate to dive the fissure where the Eurasian and American continental plates were drifting apart and, time allowing, do some whale watching. I also felt it would be helpful to have a little moral support as there was one village I still hadn't visited. That would involve asking some hard questions of the locals, and it would feel good to have my sister beside me.

In contrast to the past four months of driving alone, the car was now filled with chatter and laughter. I explained how daylight was limited and assured her that the weather could change quickly. She wasn't convinced. As we drove in the early morning traffic out of the city, she fired questions constantly. What did road signs mean? Which way was Blönduós? Were we in the Arctic Circle and could we see Surtsey? Four months of self-isolation felt like forever. Little did we know then that soon, the entire world, including Iceland, would be isolating.

I explained we were only 64° North. Blönduós was 65°, and Siglufjörður, where I had driven through the one-way tunnel, was 66° North and, therefore, officially in the invisible Arctic Circle and a qualifier for actual polar nights. The roads we drove leading to the site of Iceland's original parliament, Alþingi and Þingvellir (Thingvellir) National Park, were some of the best in the country – in contrast to what I had become used to in the north. This was not surprising, given the hundreds of tour buses that drove here daily throughout the year. I had mentioned several famous sights we would visit though I kept a few surprises for her, including a stop at Geyser. We drove the N1 ring road past the town of Hveragerði, then followed signs to Selfoss. Just before reaching Selfoss, we turned left towards Laugarvatn Lake. Kerid Crater and the Faxi waterfall were nearby, but it was still too dark for a photo stop. We drove onwards to Strokkur where, without warning, I turned into the restaurant car park, feigning a need to use the bathroom. It was too early for tour buses and the place was deserted. I persuaded Caitlin to follow me, and without saying a word, we crossed the road and walked the short path in darkness. I stood back to let her enjoy her first geyser eruption, the stench of rotten eggs she had complained of minutes before now forgotten. It never stopped exciting me how the clear magical blue bubble formed again and again before the geyser finally erupted. I had also learnt never to look away after the eruption, as on several occasions, had seen Geyser have a double or even triple eruption. My grown-up sister dissolved into shrieks of joy and laughter as, time after time, the geyser shot its plume of scalding water high into the air.

The sky was clear with not a breath of wind. Having a car afforded us the luxury of avoiding crowds, although it wasn't long before the private super jeep tours started to arrive. They had the same idea. We had enjoyed the best of the geyser and left the new arrivals to it. Next stop: Gullfoss. The crystal glacial waters from Langjökull glacier ran down the Hvita river (White River) and dropped over 100 feet into a three-tier wide curving staircase contained in a steep glacial-formed canyon. The waterfall was certainly impressive, no less so than the history surrounding the area. In the early 1900s, the owner of Gullfoss, a local sheep farmer named Tómasson, entered a contract to lease the land to Howells, an Englishmen. Howells was set on exploiting the waterfall, using its energy to drive a hydroelectric plant. Tómasson's daughter Sigríður (Tómasdóttir) sought to reduce the contract, foreseeing the potential destruction of the falls. Earning money to pay a solicitor and walking the 62 miles to the capital to fight the litigation on more than one occasion, the claim dragged on for years and became one of Iceland's most notorious legal cases. Sigríður was justifiably remembered as a formative environmentalist, and her legal representative went on to become the first president of an independent Iceland in 1944.

By the time we arrived at the golden waterfalls, so too had the first coach tours of the day. Undeterred, we walked the paths and marvelled at the sights. While Caitlin watched the vast waterfall, I realised that I hadn't seen this many people for months and became absorbed in people-watching instead. Most people looked frozen, and the few that did chat along the route thought it was windy. Haa! I disagreed.

We headed indoors and ordered a steaming bowl of meat soup, a delicious meal of salted lamb, vegetables, and barley, slow-cooked and served piping hot. We enjoyed a glorious day filled with fun and sightseeing, and I realised how having someone to share things with made experiences somehow come alive. We took a dip in the now not-so-secret lagoon of Flúðir, spontaneously stopped for a brief visit to the greenhouses of Friðheimar tomato farm and drove back to Reykjavik in darkness. I longed for an early night in my queen-size bed, though sadly, Big Sister wouldn't hear of it. The sky was clear, and the aurora forecast was good. She wanted to see the lights. "They weren't as bright as I had expected," she complained as we finally crawled into bed at well past midnight.

The following morning, the alarm clock rang, and it was the usual dash to the door to see if the Yule Lads had been. The day involved diving, and Caitlin could barely contain her excitement. Looking out of the hotel window, dark clouds were forming over the city, and white breakers rolled across Faxaflói Bay, directly in front of us. I knew it would take more than that to deter her. When she had asked me to book a dive, I chose the less famous Davíðsgjá, the darker, wilder sibling of Silfra. I had spoken to the company recommended by my local hot pot gang, and they assured me their friends in the city would give her a dream day underwater. I had hoped to book an extra dive at the hot springs of Lake Kleifarvatn, though this was not possible during winter. Our entire family adored water. It was I alone who suffered the crippling fear. I usually accompanied them to their dive spots, then waited on shore with a towel. Today was no exception, and I tried to muster enthusiasm at leaving the comfortable hotel room for a day of waiting. To Caitlin's amusement, I took my knitting. As we arrived at the site, with no hesitation, Cait changed into her diving gear in the back of the dive company van, and I admired her bravery. She and the handsome Italian guide walked the short path, and with a thumbs up in my direction, she was gone, like a seal. Her new buddy had met his match, I was sure, and he couldn't understand how she didn't feel the cold. Bluntly she told him, "You haven't dived Orkney."

Later that same day, I couldn't leave without giving Cait a tour of Althing, where Iceland's first parliament was founded in 930. More than just a historical site, it was a beautiful place to spend time. With open plains, lakes, and rivers teeming with fish, it was a stunning place and should not be overlooked. On previous occasions, I ensured I arrived before sunrise and walked through the darkness over the river and towards the picturesque church. Now it was past lunchtime, and thankfully, the tour buses had left for the day. We were almost alone. We walked towards the church, crossed the small bridge, and threw in a coin to make a wish before noticing the sign that cautioned not to throw coins into the water. I filled our water bottle from the ice-cold glacial water, taking the opportunity to splash my face and feel alert for the drive back.

Glowing with satisfaction, the drive passed quickly with talk of diving. The clarity of the water and the unique boulders, fissures, and narrow caves, not to mention a huge brown trout that had kept her company on her first dive of the day. Davíðsgjá was more weather dependent than Silfra, and as the visibility apparently wasn't optimal, they added an extra dive in the rift at Silfra also. Due to the cold weather, she and her Italian buddy had explored all four areas of Silfra in one dive before surfacing and the equipment freezing. She hadn't been so impressed, however, with the dry suits they loaned her as hers had leaked.

Christmas Day arrived, and I was glad the previous evening had been cloudy with no chance of an aurora. I lay in the comfortable bed and knitted until 2 a.m. without the need for a headlamp. Following a late breakfast, my sister and I took a brisk walk in the half-light of morning to the largest Church in Iceland, Hallgrímskirkja, in the hopes of enjoying a service. It was closed. So too was the Cathedral (Dómkirkjan), the seat of the Bishop of Iceland. Disappointed, we strolled back through the quiet streets to enjoy the rest of Christmas day in the hotel. After a sumptuous buffet dinner, we sat in front of the open gas fire and exchanged presents. Excitedly I gave Caitlin her present, which I had spent hours working on, and watched as she opened her present carefully, avoiding ripping the pretty wrapping paper. I could tell immediately that she didn't like the shawl. She had difficulty lying and eventually admitted she would never wear it. Diplomatically, she said it would be fine draped over a chair for decoration. Ha! When it came to my turn to open my present, I tore off the wrapping paper and was delighted to receive a large box of Fortnum and Masons tea and a selection of biscuits together with a new lip balm. Suddenly the telephone rang. It was the receptionist from Blönduós wishing me a Merry Christmas. We chatted for a few minutes, and she reminded me that the "Aurora Brigade" were leaving the following day, though due to the Bank Holiday, the buses were only half operating and she wondered if I could drive them to Reykjavik. Sorry no, I couldn't help them, as I was already in Reykjavik, I explained insincerely.

A few minutes later, the phone rang again. It was Laki from Grundarfjörður, who wanted to know if we could arrive a little earlier for our whale-watching tour tomorrow. The weather wasn't looking too good for later in the day, and so he would like to leave earlier than originally planned. Not for the first time, I wished Caitlin had been happy with a whale-watching tour leaving from Reykjavik, which would have been an easy 20-minute walk from the hotel along the seafront down to the old harbour. The tour boats left daily and often saw one or two humpback whales in Faxaflói Bay. Since I had told her of my trip out with Laki to see orca whales, she was determined to also take that tour. That meant a 360-km round trip over the Snæfellsnes Peninsula to Grundarfjörður. I didn't look forward to driving there but agreed we would try. Her flight home was not until the evening of 27th December.

Boxing Day morning thus saw us leave Reykjavik before breakfast and drive in pitch darkness northwards on the N1. There was no other traffic in the tunnel at Hvalfjörður, in fact, we barely met a car for the entire journey to Borgarnes. We stopped at the garage, hoping to buy something to eat but a sign on the door said they were closed for most of the day, opening at six o'clock. We drove westwards towards the Snaefellness Peninsula, and I pointed to places I had visited previously, including the farm that was famous for curing fermented shark and the farmyard where I had been forced to pitch Bothy some weeks before. This information was met with hysterics.

A light snow fell, but the road was clear. I was concerned but stayed quiet. We were both hungry when we arrived in Grundarfjörður, and so with time to spare, we made our way to the welcoming café, ordered breakfast, and collected our insulated overalls, ready for the boat trip. Laki, who arrived almost on time, explained that the group was small, with only 12 passengers, consisting of a group from Belgium, a couple on their honeymoon from Argentina, together with my sister and me. He was eager to get underway. We had a breathtaking tour with a spectacular show of killer whales. The entire time at sea was magical, and I was glad I had made the effort to drive Cait here to witness the unforgettable experience of seeing killer whales hunting, feeding, and interacting as a group. As we headed back towards the small harbour, the sky took on a stormy, sinister look, and in the short time it took to secure the boat and disembark, it had started to snow heavily. The clumps of wet flakes were getting larger due to the sudden drop in temperature and the wind blew them horizontally. This was Cait's first taste of the sudden change in Icelandic weather that I had told her about so often. "It's unbelievable," she shouted over the wind as we walked back bent double to the café. "Have you checked your weather app?" she asked me. I had checked it, and it advised a strong breeze with a chance of snow. The conditions were apparently worse up north. By the time we returned our overalls and said goodbye to the rest of the group, we heard that the mountain roads we had driven to get here were already becoming difficult to drive and would in all probability close shortly. The staff in the café agreed it was inadvisable to drive this evening without a 4×4. The last bus of the day from Grundarfjörður to Borgarnes had been cancelled, though if we could somehow get to Borgarnes, there was a bus going directly to Reykjavik. There were no reports that the main roads would close.

Our problem now was how to get to Borgarnes, which was 90 km away. There was no car hire company here. We were at a loss what to do. The café would soon be closing, and the other tourists we had met on the tour were staying locally. We were the only two people who were stranded. I was sure we could find accommodation for the night, and if all else failed, I'd persuade Caitlin to sleep in the tent with me, but for now, I remained quiet on that subject. I was more concerned that her flight was leaving from Reykjavik the following evening, and there was no guarantee that the roads would be passable early enough in the morning to drive her there. Instead, I chose to do what I had done often in Iceland. We left the warmth of the café and followed signs for the swimming pool. I asked Cait, did she feel like a hot pot? And in return, she asked if I was joking. The young lady working on reception greeted us like old friends and happily found Cait a swimsuit to rent. Hilariously it had two giant pineapples printed on the front and given that my sister was rather well-endowed in the boobs department, I couldn't stop laughing. I was glad I had my own costume with me. I quickly went on to explain our dilemma to the young lady, making sure to let her know I wasn't simply a tourist but lived temporarily in Blönduós. That seemed to hit a note as she explained her grandmother was born there and that she stayed with her every summer. We made our way to the changing area, showered, and joined the locals in the small pool. Cait had followed less enthusiastically with her hands folded across her chest. It didn't help when the only two words that were directed at her from an elderly lady were, "Pina Colada!"

A quarter of an hour hadn't passed when an extremely tall young man dressed in red speedos came to the side of the pool and said in broken English that we had to get out of the pool. I asked why, and he explained that if we left immediately, he was willing to try and drive us to Borgarnes if we were willing to pay him. By the time we threw our clothes back on and returned the rented fruity swimsuit, our driver was already in the car park with the engine of his super jeep running. He talked little, and we allowed him to concentrate on driving. Visibility was nil, and I was glad for once that I was not the driver. Sadly for Cait, she saw none of the spectacular scenery that the area offered. It had been dark when we arrived, and we now sat looking out of car windows that were fogged up against a backdrop of driving snow that, on a clear day, would have been a photographer's dream. It was famous for being the setting of Jules Verne's *Journey to the Centre of the Earth*. Caitlin agreed that she would have to come back another time to see what she had missed on this trip.

Finally, we were unceremoniously dropped off on the main road on the outskirts of Borgarnes and had no choice but to start walking in the direction we hoped the bus might come from. We had checked the online timetable and knew we had missed the bus by a few minutes from the bus station. Our chauffeur apologised that he could do no more for us. If he didn't turn around immediately, he might have been stranded himself. He told us that he had a Tinder date that evening with a tourist from Belgium, which he didn't want to miss. Our gamble paid off when we saw the welcoming sight of a yellow and blue Strætó bus speeding in our direction. I jumped and waved and ran in front of it to force the driver to stop. Amazingly, he did so and even shared a smile as he opened the door and welcomed us on board.

"I'm starting to love Iceland," said Cait as we settled ourselves into the heated seats and fastened our seat belts.

From Borgarnes, it would have taken only half an hour to drive to the village of Kleppjárnsreykir. My original plan to visit the village in company would now have to wait until another time. For now, I was relieved that we had caught the bus and Cait would get to the airport in time for her flight home the following day. I didn't yet know when or how I would be reunited with my Volvo. In our haste to leave the pool, I'd forgotten that I had the shopping the ladies in Blönduós had asked me to buy for them, which was still in the boot of the car.

The following afternoon we said our tearful goodbyes. Caitlin tried her best to persuade me to call it a day and leave Iceland. Hah! The bus left from our hotel for Keflavik International Airport with my sister on board. I said goodbye to the luxury of the hotel and made my way to the BSI bus terminal to catch the next bus to Blönduós. I rang ahead to the N1 garage and diner to ask if they could save me a portion of meat soup for dinner.

THE SITUATION

By now, most people in the small town recognised me. I had lived in their community for four months and was happy to be referred to as "the Scottish woman". Slowly, people had become less suspicious and more welcoming. The older generation curiously asked questions about "The Americans" and how were the living conditions in the "School". They had noticed the increasing variety of unpronounceable and overpriced food items in the local supermarket and thought these visitors were mad spending so much money on "shite". Manuka honey, Matcha tea, avocadoes, and Black Beluga lentils were all unnecessary items as far as they were concerned. Several people asked me how much the artists charged for their "work", imagining it had to pay well. Being handy themselves, and with long winters stretching ahead, they wondered, could they too sell their handicrafts and knitting in America?

I loved that Blönduós was opening up to me, and I felt more and more at home there. When people shared stories with me, I asked if I could take notes, and they were happy for me to do this. They happily invited me home for tea and cake, with their grandchildren often acting as translators. I was privileged that they felt comfortable enough to share their photo albums and family histories with me. News in the small town had spread that I was interested in writing, and suddenly, everyone had a story.

I was on my way into the swimming pool when I greeted Hulda and her husband – the pair who had kindly loaned me their car to visit the grotto while Hulda had enjoyed her goose party. On our journey north a few weeks ago, she had asked me, had I heard of the "Situation" or, more precisely, "The Children of the

Situation"? I replied that I hadn't, and she went on to explain in some detail the events that had taken place many years ago. She insisted I meet her grandparents, who could tell me much more as they had personally been affected by The Situation.

During the early 1940s, the allied forces occupied Iceland. By all accounts, it was generally a peaceful occupation. The population of Iceland at that time was around 120000. Suddenly, at the peak of the occupation, around 25000 young British soldiers were posted to the island. In turn, they were replaced with around 50000 American servicemen.

It wasn't surprising that romances developed between the young, handsome, and well-mannered soldiers and the fresh-faced, innocent Icelandic girls. Iceland had been relatively sheltered from the outside world until then. Hundreds of girls courted the soldiers, and many went on to marry and have children. Others faced pregnancy alone, their babies fatherless, and their community branding them bastards forever. These children were referred to by the Icelandic authorities as "Children of the Situation" or in Icelandic, *Ástandið*.

The relationships were natural and consensual. Sadly, however, Iceland condemned the women's behaviour as promiscuous and nothing less than prostitution. The Situation became so critical that the government banned all women from having relationships with soldiers. The local men, including fathers, uncles, brothers, and friends were encouraged to report the "whores" in their community and the punishment for disobeying the rules was severe. In less than two months, almost 500 girls' names were gathered and recorded as being under suspicion.

Hilda's family lived in the town of Kleppjárnsreykir. This town housed one of two reformation houses where the girls were sent to, to serve out their imprisonment. This was no easy sentence with the girls, some as young as 12 and the eldest recorded as 61 years old enduring inhumane conditions and solitary confinement. Degradingly, their virginity was tested by a doctor to confirm whether they had been sexually active. The women's word was not enough.

Despite these draconian measures, some relationships did survive, and many women later left Iceland to start a new life in America or Britain. Iceland has never welcomed the women or their children home. No pardon or apology has ever been offered, and despite most of these women being able to trace their roots back several generations in Iceland, they have never received a public admission of wrongdoing for the abominable persecution.

The two houses accommodating the "Yankee whores", as they were called, were open for a full year and were permanently closed in 1943. This piece of Icelandic history piqued my interest, and I was desperate to learn more about the lives affected and to verify facts that I had heard second- and third-hand. I wanted to visit Hulda's hometown and meet the survivors who still had memories of those grim times. Before leaving Iceland, I did interview four families directly linked to The Situation and was captivated by their shocking tales of cruelty intertwined with the beauty of their love stories. Just like my mother before me, I filled my notebooks as they spoke, and knew this was something I would research further in the future. If I needed a reason to return to Iceland, I had now found it.

Hulda promptly asked if I had visited her grandparent's town and was disappointed when I told her I hadn't due to the weather. "It's not the weather that's the problem, it's your car. It's shit," she replied. I asked if we could plan a trip together, and she replied, there was no need as they would be travelling to Blönduós for Þorrablót, and I would meet them then. I had no idea what Þorrablót was but looked forward to it anyway.

My hosts for New Year, a lovely couple I had met and chatted to several times, waved and came to join me in the hot pot. She was a beautiful, talented, entrepreneurial lady who seemed to be good at everything she turned her hand to. Her husband ran a business also and was dedicated to keeping his waist trim by using the gym daily without fail. They had apparently walked past the old school to see if I was at home, but as they had not seen my Volvo, had assumed I was still in Reykjavik. I explained that I had been forced to abandon my car at Grundarfjörður. "Why didn't you leave your keys there?" they asked, surprised at my stupidity. They could have easily found someone to drive it north for me. I hadn't thought of that at the time, I explained. Rapidly they hatched a plan and made a call. Their daughter's boyfriend, whom I had never met, was unemployed and had time to drive me to Grundarfjörður the following morning. The weather had eased, and the roads were clear. If we left early, we would be back in time for the New Year's party that I had been looking forward to.

New Year was a huge celebration in Iceland and would not be complete without fireworks. The pool got busier, and everyone seemed to talk about fireworks. For my benefit, they explained that the sale of fireworks was restricted to ICESAR, who had the license to sell them for a limited period, usually between 28th December and 6th January, and this was a valuable portion of their annual income. The latest year saw a whopping 800 million isk (around 4.5 million pounds) spent on fireworks in Iceland. I wasn't interested in buying any myself, but my offer of a donation towards the evening's cost was promptly declined.

So, the following morning, bright and early, I waited beside the N1 garage for my lift to Grundarfjörður with Arnar. The bright pink pick-up was a surprise. So was Arnar's taste in music. Willing myself the entire journey to be grateful, I wondered, was there a secret camera filming me? I tried to appear interested in the truck, but all I learned was that she was the "Pink Lady". As for the music, which was played at excruciating volume, I could honestly say that I had no idea what the hell he talked about when he explained this was electro-techno trip hop. Apparently, I was listening to haunting whimsical beats that were inspired by and worked in synchronicity with the mysterious landscape. Personally, I thought it was an assault on the ears.

Glad to arrive in Grundarfjörður, I thanked Arnar and offered to pay for his fuel. He readily accepted and thought maybe I could give him something for his time too. We negotiated, and finally, I waved Arnar and the "Pink Lady" goodbye. Peace at last, I thought as I set to scraping ice and snow from my car.

I wasted no time in driving back to Blönduós. I didn't want to be late for the evening's festivities.

I COULD EAT A HORSE

The last day of the year was quiet in Blönduós until around five o'clock when everyone seemed to be coming or going. The residency was almost empty, many artists having headed back to the US in time for Thanksgiving and Christmas. The Aurora Brigade weren't missed, and there were just three artists remaining, plus myself. I decided to invite the trio to join me to watch the town's bonfire later that evening. I had been invited to spend the evening with some locals I had gotten to know, but I was sure we all would be heading into town to watch the traditional bonfire followed by fireworks. "Too cold," said the two girls, while the only male artist, a middle-aged American, had a date with a man he had met earlier that day on Grindr. It sounded as though Internet dating had taken off in Iceland.

I wore the only dress I had brought with me, paired with unflattering thermal tights and my heavy Lundhags Boots, as I guessed the evening would include a lot of standing outside in the cold. I left my car and instead decided to walk over the bridge to the other side of town. Not for the first time, I cursed the blasted boots when, halfway up the hill, I slipped on the icy pavement and ripped a hole in my tights. I also managed to pull half of the arm off my dress as I fell, which meant that I arrived a few minutes later looking somewhat dishevelled. I apologised and asked if I could use their bathroom to tidy myself up. Feeling uncomfortable at the best of times when being introduced to new faces, I felt mortified as I tried my best to hide the holes in my clothing. Having taken a bottle of champagne as a gift, I was glad they opened it instantly. It was sometime later before I realised no-one else had touched it. Apparently, they thought I had brought it for myself rather than drink the wine they had provided for their guests.

I hadn't eaten since breakfast, and dinner smelt delicious. "I could eat a horse," I said to Arnar, who now sat next to me.

"Yes, soon," he replied.

I laughed politely at his sense of humour. Therefore, I was surprised we took our seats at the beautifully arranged dining table to be served fillet of foal served medium rare with all the trimmings of a traditional roast dinner. It was the first time I had eaten horsemeat, and it wasn't bad at all. I asked jokingly what this one's name was. The entire table fell silent as my host explained they would never eat a horse with a name, as it would be like eating a family pet. Not for the first time that evening, I felt embarrassed. I was thankful when it was time to once again dress for the outdoors and walk into town to watch the bonfire being lit. I stiffly walked back down the hill, praying I wouldn't fall over for a second time that evening, while enviously watching the other ladies racing ahead in their high-heeled shoes.

As a tourist, joining in with the locals to watch the traditional bonfire at New Year provided a great opportunity to enjoy one of the country's oldest traditions. Built by the local council and volunteers, the bonfire was huge, and once lit, was spectacular. The entire community had come out to see it. I was told that it was polite to mingle, so walked around wishing everyone that made eye contact a "Happy New Year". Some were friendly and wished me too, *Gleðilegt nýtt ár* (Happy New Year) while others looked away a little embarrassed, probably having heard I was mad. Feeling self-conscious, and as it was only a short walk from the residency, I rang the two girls and again asked, did they wish to join me? But they were happy watching the fire from their bedroom windows.

The entire gang from Blanda Search and Rescue were working, making sure the fire was burning safely and ensuring the evening went smoothly. One of the men in the group asked why I hadn't been in touch to accompany them on a call-out, and I promised I would do so the following week. I still felt embarrassed about my antics on the cliffs where I had injured myself and almost fallen to my death. In such a small village, it was no good trying to hide. At nine o'clock, we regrouped and again walked over the bridge, past the hospital and police station, the winter home of Birna the Polar Bear, and up the hill back to the warmth of the house to "welcome in" the New Year. Glasses were filled, and at exactly 22.30, the music was replaced with the TV, and everyone sat down to watch Áramótaskaupið (The New Year's Lampoon). Since it started in 1966, this had become a tradition in Iceland, a typical satirical take on the previous year's news, politics, and activists.

At a few minutes to midnight, without warning, everyone stood up, again dressed in their outdoor wear, and with a fresh drink in hand, headed outside to light the fireworks. For such a small town, I was astonished how impressive the display was. I could easily believe how ICESAR could earn so much money from the sale of the fireworks. At midnight on the dot, to a popping colourful crescendo overhead, a toast was made to old friends and new. I thanked my generous hosts for the evening and made my way once again back along the slippery pavements and in through the solid wooden front door of the residency. I didn't have a lump of coal but instead placed an old silver sixpence in a corner, under the stairs, as a first footing gift for the building. It had been the perfect way to welcome in the New Year.

It felt like no sooner had the Yule Lads headed back to their Dimmuborgir Mountain lair and the Christmas excesses had worn off than the town was preparing for the next feast, that of mid-winter. As usual, I heard about it first in the pool.

"Thorawhat?" I asked.

"You won't like it," warned a local named Barney.

"It's shit," said Ragna when I asked her for more details.

The selection of food in Icelandic supermarkets could be a challenge at the best of times and once Christmas had been "reduced to clear", it was quickly replaced with some decidedly dodgy-looking items. Why? Because between late January and early February, Iceland celebrated Þorrablót. The name was derived from January in Old Norse (*Thorri or Þorri*) and blot (meaning sacrifice). In days gone by, a toast was made to the Norse gods. I liked to think of myself as an enthusiastic foodie and thought that this would be a great opportunity to taste some more Icelandic delicacies. I was, therefore, delighted to receive an invitation to Hulda's house to eat some genuine Icelandic food. It would also be an opportunity to finally meet her grandparents. What I hadn't reckoned on was seeing her guests dressed in their finest clothes, tucking into food with their bare hands, bringing half a sheep's face to theirs and relishing it with finger-licking enthusiasm, which extended to consuming the tongue, eyeballs, and ears of the beast. The experience was worse than my first Swedish crayfish party, where my aristocratic hosts sucked and slurped noisily to extract the juices and brains from the dill-saturated crustacea. It would take some time before I got over this midwinter feast where I watched horrified as they ate the head of a sheep, the ram's balls (literally), links of horse sausages, and an array of sour meats in aspic jelly – all greedily devoured. A party wouldn't be a party without a side order of fermented Greenland Shark (*Hákarl*) or dried sour whale (*Súr Hvalur*). These delicacies were washed down with dark beer and a heck of a lot of Brennivín. My hostess Hulda happily accepted a packet of Swedish Knäckerbröd (crisp bread) and some jars of Sill from Ikea, though they remained unopened.

The strange dried, frozen, and reformed foodstuffs that the Icelanders love to eat during the feast of Þorrablót could be found in all food shops. It wasn't always easy to know what you were buying in Icelandic food stores, and on several occasions, I had made expensive mistakes. I would open a packet only to promptly close it again and throw it in the bin, unsure what it was but certain I didn't want to eat it. If you couldn't read the label, beware! It suddenly made dried fish, which resembled something found on the floor of the spa after a pedicure, rather less distasteful.

There was no time to both eat and talk during the feast, so I made a promise to drive over to Kleppjárnsreykir as soon as I could and enjoy tea with Hulda's grandparents. I got home late that evening, and as I quietly closed the fire door, I felt the past ladies of the school looking down at me from their portraits. Despite many setbacks, this school has remained open and the portraits along the corridor have witnessed many goings-on. I wondered what they would have made of the recent Grindr date. The male artist, a teacher of textile art and an enthusiastic quilter, had arrived two weeks beforehand and entertained us all with his uncomplicated fun character. The day he arrived, I was envious of his natural confidence, in contrast to my own arrival, which had been weak and teary. He took the stairs two at a time, dumped his elegant luggage outside his bedroom door and headed out directly to see what the town "had to offer". I'd been emptying the washing machine at the time, and from the upper window, watched him walk purposefully, dressed from head to toe in the latest New York fall/winter collection, everything immaculately polished and pressed and with a large leather man bag thrown elegantly over his shoulder. I couldn't imagine a more opportune time to breathe the phrase "You aren't from 'round here, are you, Boy?" and wondered what the locals would make of him. He was a warm, gifted, and genuine man who, instead of searching for some deep inspiration, hoped only to find a little love in Iceland.

The school has undergone and survived many transformations and tribulations. Apparently, as far back as the First World War, a boat had capsized off the coast of Blönduós. Both during and immediately after the war, the amount of coal brought to Iceland was very scarce, and it was a struggle to find wood to burn, given the limited number of trees on the island. Schools closed due to lack of heating, and yet fortuitously, the shipwreck and the healthy plundering of its timbers by the female students ensured that this school remained open. A fire in the school some years later threatened to destroy the building, yet thanks to some elf karma, it didn't catch hold despite the perfect conditions of it being built like a tinderbox. I thought back to the night of the storm and was sure the building had seen off worse.

There was an unusual calmness in the building, and when there was no-one else about, I felt at home there. After an interesting evening, having drunk too much aquavit and eaten too few Þorrablót delicacies, I closed and locked the front door, climbed the stairs, and made my way along the corridor, past my own bedroom door to the kitchen and emptied my pockets of half-eaten smelly food in the bin. I bid the portraits good night and closed my bedroom door, and fell asleep counting sheep heads.

BLESS BLESS BLÖNDUÓS

As the weeks passed and the daylight lengthened, my thoughts turned to home. I could hardly believe I had achieved my original goal of staying 21 weeks in Iceland. The ferry from Seyðisfjörður was due to leave in less than a week, and I had confirmed my booking. I had no reason to imagine the drive to catch the ferry would be easy. The past week had seen some sunshine, and I had made the most of it. Tempted though I was to drive to the ferry port in good weather, it would have been a long wait, and I still had one area I wished to return to. I loved the freedom of having my tent and decided that one final road trip was possible before packing up and heading eastwards and home.

With the car fuelled, the remainder of my groceries packed, and my grab bag on the front seat, I drove out of Blönduós and headed south. I picked up the road sign to Hólmavík and followed it west and onwards to Súðavík. Disappointingly, the tourist information centre was closed despite the sign saying "Open" on the door. The last time I had driven past, it had also been closed. I imagined they rarely saw tourists during winter; I hadn't seen a soul for most of the journey. I had hoped they could give me a few tips as to where to camp. The road conditions were horrible, and I concentrated hard to avoid potholes and verges but was thankful that the good weather stayed with me. The sun might have been shining, but the wind chill was ice-cold. Occasionally stopping for a photograph, I gave up on opening the car door to step outside, instead opening only the window enough to balance the camera for a quick shot.

I finally reached Ísafjörður, the jumping-off point for Hornstrandir, late in the evening. I had visited this area only once before in the hopes of spotting an arctic fox in its natural habitat. The drive was a difficult one, given the poor quality of the unpaved roads, but the scenery through the fjords and mountains was worth the effort. The gloriously wild Nature Reserve of Hornstrandir provided a safe habitat for Iceland's only native mammal. There were no roads through the peninsula, so access was limited to a boat or a long hike from the mainland. The two nights I stayed in the area, I sadly didn't see a fox. I had started to feel unwell before leaving Blönduós, and by the time I spent the first night in my tent, I had a full-blown cold, my entire body ached, and I shivered with fever. I drank hot blackcurrant juice and ate honey sandwiches and a tin of sardines. I tasted nothing but knew I needed to eat. I took painkillers for only the second time on the trip, despite being in pain from my foot injury. I had taken two after I had a narrow escape on the cliffs at Hvítserkur, otherwise had not used any medicine. I wondered, as I was nearing the end of my Icelandic odyssey, whether I was subconsciously worrying about leaving Iceland and what the future held for me. I knew I had come a long way on my emotional journey and had dealt with much of what caused me pain and anxiety. I had detached myself from my old life, made the necessary changes required to move on unencumbered and yet knew that the road ahead once I returned home would require continued efforts to ensure I didn't relapse. I had achieved what I had set out to do. I hadn't given up, and Iceland had rewarded me on many levels. For the first time, having pitched Bothy, I had little desire to sleep in her. Had there been a hotel nearby, I would have gladly checked in. Sadly there wasn't, which meant a miserable two nights nursing my cold before packing up my final camping trip. The knitted square remained unfinished though I vowed to complete it as soon as I felt better.

Back in Blönduós, I spent my final days in a flurry of packing, cleaning, and following up invitations that I had received. I took tea with Ragna outside to avoid an allergic reaction from Kopi the parrot. I knitted with the ladies in our knitting coffee circle. I gifted my unwanted items, including an Italian coffee machine and some crockery, to the residency kitchen. The shopping that I had done in Reykjavik over Christmas, which had remained in the boot of my car in Grundarfjörður, still sat in its carrier bag. The artists had left for home without paying for it. It hadn't cost too much, and I would likely never use tapestry threads or sewing machine oil, so I handed them into the local charity shop, where I knew they would find a worthy home. The bedside table and lamp I had bought in Ikea would remain in the bedroom for the benefit of future visitors.

I drove to the garage next door to Istex and retrieved my summer tyres to pack in the car before the lighter items could be placed on top. The boys in the garage insisted on washing my car as a favour and told me to return in summer for Húnavaka. I had no idea what Húnavaka was but promised I would try anyway. Life continued as usual in the residency, and arguments broke out between the remaining few residents as to who would move into my room when I vacated it. I chose not to involve myself in their petty squabbles. My last dip in the pool was no different from any other I had taken. The steam room was still broken, and the locals chatted among themselves. I sat in one corner of the hot pot, trying desperately to avoid foot contact with the other bathers. On leaving, I thanked the staff for everything they had done for me and gave them my tin of biscuits which I hadn't opened since receiving them at Christmas. I walked home for the last time in the dark, noting that the Kill House still stank despite its reduced workload. Blönduós hadn't suddenly become beautiful. Many people still ignored me. The rusty containers and discarded machine parts lay where they always had, and still no one seemed inclined to pick up litter from the beach. Despite all that, I felt a connection to the town even though I didn't love it. Blönduós was where I had lost Mats. I felt I had, in some way, earned my own rite of passage too while staying in this unpronounceable rural backwater. Still, when I spoke to people and explained where I was, they mocked, "Where the fuck is Blönduós?" It was with mixed emotions, when the final day arrived and with no one to wave me goodbye, my handbag on the seat beside me, my car packed, and my jaw only slightly clenched, I drove out of town for the last time, turning left at the N1 garage for my final road trip east. Bless Bless Blönduós.

I slowly drove the familiar route east to Akureyri, drinking in as many memories as I could. Making my way through the town for the last time, I noted that no longer were the abundant rowan trees heavy with fruit as they had been in autumn. They grew well here and could trace their origins back to folklore where they were worshipped. It was supposedly on the site at Modrufellshraun where a brother and sister were executed for incest, despite their claims of innocence. A rowan tree had grown up from the blood spilt that day, and despite many attempts to kill the tree, it always grew back and flourished. There was much history surrounding this trading town with prominent names of merchants, clergy, professionals, and politicians kept alive through street names and districts. Akureyri was a poor town, though thanks to donations from wealthy merchants and even Danish settlers, a hospital, printing press, botanic gardens, and industries were built, offering a healthy infrastructure from which the town thrived. During the early settlement years, there was an ancient assembly site here, located in Oddeyri, where in 1550, the last Catholic Bishop of Iceland and his sons were pronounced guilty of treason and executed. The first house was built in the town around 1795, where a trading post operated uninterrupted until 1933. The first woman to vote in Iceland lived in this northern town, and she was also the first woman to seek a divorce in Akureyri.

I crossed the bridge over the Eyjafjörður fjord and watched a plane attempt a windy landing on the airport's runway to my right. I had been tempted to stop at the small garage and say goodbye to Ragna's brother, but it looked busy, so I decided to keep driving up the mountainside and east towards Mývatn. The weather was grey, though the wind stayed under 10 m/s, for which I was grateful. Sadly, I saw no whales surface today. Allowing myself ample time for the ferry, I would break the journey and spend one night in the Mývatn area. Three nights' sleep left in Iceland before M/V Norröna left for Denmark. There were still places I had yet to visit, but on the whole, my map on the seat beside me was well highlighted. Planning this final journey too far ahead was a waste of time, given the weather, and I preferred to take these last days as they came. Mývatn was a beautiful area, not only famous for its birdlife and Yule Lads, but it was also home to one of the country's leading nature baths, and my swimming costume was in my daypack ready. I hoped to enjoy an arctic char sandwich and some homemade pie at the Bird Museum.

Events in our own lives formed us, while world events could quickly influence how we live. The past six months had allowed me to realise my dreams, and deep in my heart, I had achieved what I hoped I would. Surtsey and its unconscious reminiscences were behind me now. The past was buried, and so too was my best friend. I couldn't imagine a future that didn't include Iceland and planned to return in summertime to continue where I had left off, researching future projects. Firstly, however, I would return home to Sweden, finalise the purchase of a small property and make a home for myself. Little did I know that it would be more than two years before I would return to Iceland, the entire world would be in lockdown directly or indirectly, and travel would unlikely be the same again.

The dramatic winter landscape stretched before me, and at first sight, seemed sterile. As I had found over the entire land, to look closer revealed a bounty of plant life, mosses, lichens, and fungi. Wild angelica, miniature wild strawberries, and bilberries grew in abundance. I had, over the months, taken photographs of plants I'd never seen before, and on my trips to Akureyri, had visited the Botanical Gardens and museum to verify their names. Several I had found were rare and found only in certain rocky terrains. Field Scabiosa and Stone Brambles, Cowberry and Red Sorrel, Arctic Thyme, and the beautiful national flower of Iceland, the Mountain Avens, all appeared before my feet, and all seemed to flourish. Iceland wasn't just about clocking up the miles or ticking boxes. So often I heard of tourists allowing themselves two or three days to "do" Iceland and felt their achievements missed the mark.

There was so much more to Iceland than Reykjavik and rotten shark meat. There was the famous Blue Lagoon, the charming, if bumpy, Ring Road with its hot springs and waterfalls aplenty. The whale-watching was world-class. The food was interesting and often delicious. There was natural energy and seismic activity, and if you waited long enough, you could witness all manner of natural phenomena, including geysers, volcanoes, hot springs, rainbows, and the northern lights. The weather was changeable, schizophrenic at times, causing avalanches and mudslides, white outs, and flooding. The wind was so significant it had more than 150 words in Icelandic to describe it. By comparison, snow had a paltry 25!

It was an undeniable fact that Iceland was a nation of survivors. Crisis after crisis could have seen them off, but Icelanders were made of sterner stuff. Eruptions from deep in the bowels of the earth, glacial winds causing blinding storms of dust, thousands of earthquakes in a single day, fog as dense as milk, and snowstorms which rendered you breathless were taken in a stride. If that weren't enough, be careful where you stepped as you would come across boiling water and mud spouting from the not-so-terra firma, flooding from glacial run-off, cooled lava as sharp as razors, crevasses tens of metres deep, and volcanic ash that could swallow a person all added to its dangers. How could anyone possibly see, feel, hear, smell, or taste Iceland in three days?

On a stretch of the river Laxa, the most famous salmon fishing river in the country, I had some weeks before I noticed a beautiful hotel with stunning views. I had called ahead and managed to get a knock-down rate for bed and breakfast. I arrived at a quarter to four in the afternoon, and it was already dark. It would be a long night reading as I wasn't tired and couldn't muster the enthusiasm for a walk in the pitch dark. The icy car park had been treacherous enough. However, I had packed my knitting.

The hotel mainly catered for group travel, and they were curious to welcome a solo traveller at this time of year. On first sight, the wraparound wooden decking would be a glorious place to pull out a chair and watch for the lights. On closer inspection, it was a skating rink and wind tunnel and more suited to bird watching in the summertime. The aurora forecast was poor as the cloud cover obliterated any view of the stars. The hotel was void of customers. Had I been paying full price, I'd have unpacked Bothy and cancelled my booking; however, they had offered me a special winter rate, and I really couldn't face a night camping.

The room was spacious enough, and the bathroom glorious in comparison to that which I had endured for the past few months at the residency. With its sparkling white tiles, a power shower, and several fluffy white towels, I couldn't wait to cast off my heavy clothes and enjoy some luxury. With little to dislike, and despite the overhanging shelf over the bed catching my head as I woke up, giving me a bruise the size of a plum on my head for more than a week, I couldn't fault the room. The notice beside the bed read that reception was happy to provide a free wakeup call if the "lights" appeared, saving guests from waking up every hour to check for themselves. It was a nice touch. I slept like a log, and the following morning, after a limited and uninteresting breakfast, I checked out and drove the short distance to walk the Krafla caldera. The information in the hotel said it was an easy walk, even in winter, and I was looking forward to stretching my legs. My foot injury was now causing me considerable pain, but I could do nothing about that until I arrived home.

The paths were quite steep in parts, and without good footwear, I wasn't sure I'd have continued. The weather had changed for the worse since leaving Blönduós the day before, and the views were non-existent.

In such a vast wilderness, it was understandable that people believed in hidden folk or Huldufólk. Tracks in the fresh snow around me showed I wasn't alone. No human prints but small steps in straight lines. Foxes. Their trails were everywhere. Haa! I'd keep my eyes open, as in the grey morning light, I imagined they'd be chancing a meal of a discarded bird egg, or a lemming, and I still hadn't seen a true arctic fox other than two which were in a cage outside the information centre at Súðavík, which in my opinion, didn't count.

I headed back to the car and drove onwards to Dimmuborgir, the home of the Yule Lads. The visitor centre was open but turned out to be a disappointment. I was desperate to pee and headed for the outside loos, which had a pay-before-you-pee system. I had no cash, only a card, so shuffled inside to debit a few kroner from the machine and headed back across the slippery walkway to finally relieve myself. In compensation, the natural lava formations here, however, didn't disappoint. It was so deformed it was scarcely believable.

THE FINAL LEG

I heard once that we never truly know ourselves until we spend time hungry, cold, afraid, and alone. I had by now experienced all these things and was ready for this adventure to be over.

The sky was moody, and the temperature was dropping fast. Ahead of me still was the worst part of the journey, the 200 km drive across the mountain that I had dreaded since the day of arrival. I could put it off no longer. I checked the weather app, and of course, it said the bad weather would become worse. Why wouldn't it?! There had been significant snowfall over the past weeks, especially on the high ground, and although my winter tyres gave good traction, it was the thick snow that caused me the most problems. My car was so low to the ground that it literally ploughed the road as I slowly drove forwards. The wind had blown the fresh snow into drifts, and the higher the terrain, the worse it became. All I could do was cross my fingers and press on.

I had driven thousands of kilometres since arriving and experienced blind summits, one-lane tunnels and small bridges, poor road markings, potholes and much more. All of these were only exacerbated by inclement weather. Ice, hail, rain, snow, and wind beat the hell out of the island with regularity, and all served in making driving difficult. Despite that, it appeared as though every kilometre of this final journey would call for heightened concentration, nerves of steel, with shoulders raised towards my ears. When I had first driven this road, I had climbed. Now it was downhill all the way. As I passed signs cautioning speed I laughed nervously: I could barely do 30, let alone 60 or even 90 mph on these roads. Traffic was light, and the little that did share the road was four-wheel drive with extra high chassis and huge tyres. I had once imagined that these types of cars were built only for the interior. Now I knew that the majority of northern Iceland was the interior.

On and off, it snowed, yet the roads remained open. Many road signs were impossible to read due to a covering of snow, and those I could read, often advised 4×4 vehicles only or road impassable. So far, the road to Seyðisfjörður remained open. I stopped from time to time to check the car, used my shovel to remove a build-up of snow and ice on my wheels and under my car, and more than once to remove a snowdrift that had recently formed. I had to get to the ferry today, come hell or high water. By the time I arrived at Egilsstaðir, I had almost lost my nerve to continue. The petrol station was open, and I went in for the company of people more than anything else. I asked if anyone had news of the ferry, and they confirmed it had arrived. It would be leaving tomorrow as planned.

A man with his two sons spoke to me and was interested in hearing I had stayed in Blönduós for six months. His mother had been a student at the original lady's school. He recommended I washed my car and took a bag of chamois leather cloths from the boot of his own truck to dry it for me. I think he craved some company, and for that moment, so did I. I hoped my door sills wouldn't freeze overnight and tried to remember to grease them with silicon when I parked for the evening. On saying goodbye and thanks, I introduced myself by name. He offered a firm handshake and a smile. His name was Jon, without an H. I, by now, had realised this was a common name. Hah!

After a further check of the road app, the weather conditions were deteriorating fast. The roads I had travelled earlier that morning were now showing red, which meant impassable. The sky spoke for itself. I was ready for my nemesis, the mountain, so got into the car and faced my fear.

The last 30-minute journey from Egilsstaðir to Seyðisfjörður took me an hour and a half. The road was glass, and the gradient was alarming. It had, however, stopped snowing. The road conditions might have been abominable, but the worst weather was a short distance behind me. I had to keep going before it caught up. I tried to keep my foot off the brake but, at times, lost my courage. When the car sped up involuntarily, I lightly pumped the brakes. When I braked, I skidded; when I skidded, I stopped – thankfully before going over the edge of the mountain – and when I drove again, I was back to accelerating without my foot on the pedal. The hairpin bends and alarming drops were suicidal. I tried to stay in the middle of the road as best I could. A wild reindeer looked on, unimpressed. Finally, the mountains opened ahead of me, and I saw the miniature town before me. I also saw the last snake of road I had to drive to reach it.

Metre by metre, I gently manoeuvred the car in the direction I hoped it would take. Finally, with a huge sigh of relief, I saw the little periwinkle blue church and the huge car ferry docked in the harbour. My car and I had made it.

I checked into the guesthouse, then headed to a small café at the junction of the town centre, lit the candle on the table in front of me and wrote my final postcards. It was an old-fashioned gesture but one I still enjoyed. There was little point in checking the weather app now, but out of habit, I found myself doing it anyway. The roads I had just driven between Mývatn, Egilsstaðir and onwards to Seyðisfjörður were now closed. The unpredicted hurricane, which hit land that evening before sailing, immobilised the local power supply to the mains water, thus leaving the town without running water or flushing toilets for a day. *Perfect timing*, I thought to myself.

My winter in Iceland had come and gone, and it was now time to head home. I had had my fair share of winter weather, with its constant storms, winds, blizzards, darkness, and cold, and now, I looked forward to spring.

The past months had taught me a lot about weather, and I had especially learned a new respect for the wind. Its violence, bordering on psychopathic, could cause untold damage as it crossed land. At sea, I knew it could be much worse.

Since my childhood memories of the lifeboat disaster, I had hung on to my fear of deep water like a safety blanket. Avoidance was best at all costs. As I sat in the small café, half interestedly looking at the weather forecast, I suddenly sat upright and swallowed hard. An extremely violent storm would meet the ship somewhere between Tórshavn and the west coast of Norway. I felt sick. In the winter, Norröna morphs from a passenger cruise ship into a container ship with limited crew and few passengers. I should have given this vital detail more thought. The temperamental winter storms had a lot to do with that. No amount of ruminating over satellite imagery changed the outcome. There was bad weather ahead, and I was about to board the ship.

In the bitter cold and inky black night, I left Seyðisfjörður. No sooner was the ship underway than I felt the swell, gentle at first but perceptibly there. I glanced backwards over the mountains and would never forget the spectacle. The most glorious aurora waved me *bless* (goodbye in Icelandic). The colours were a spectrum of green to gold to rhubarb red, streaked with azure blue and platinum ribbons. The camera was useless, my hands shook with cold, and the ship's movement blurred the images. The ship was underway, heading into a storm, and I was trapped. Mats was no longer there tracking my journey. This time, I was truly alone. The predicted sailing time was 70 hours, and despite all that Iceland had thrown at me, these would be the longest of my life.

As the vivid aurora paled and the silhouette of land melted into the night, I comforted myself with a warm shower. It had been a long day; in fact, it had been a long six months with my Icelandic adventure punctuated by extremes. I tried to sleep. The hours rolled by, and I counted them. The swells were there though not alarming. We were a third of the way on the map, and I prayed the forecast was wrong. Covering the nautical miles through the Faroese Archipelago, the sky turned sinister and a towering form of vertical cumulus puffed up over the rugged mountains. Northern Fulmars dived and bolted towards their cliff top seclusion as though battening down the hatches. Perhaps they knew something too. Less than 20 passengers were on board a ship that could easily carry 1500. My car was one of seven, whereas in high season, she could accommodate 800 vehicles. Wandering from one deck to another, I felt as though I were on the Marie Celeste.

Leaving Tórshavn, we motored out through the watery tunnel of islands, leaving Streymoy behind us. Our course was south towards Norway and on to Denmark, and the swell became more noticeable. By dinner time, walking to the restaurant included a simulated drunken gait. I felt the first adrenalin fuelled fear course through my veins. As though by the minute the sea gained velocity and staring out at the grey-green watery abyss and high waves, I knew I had to accept whatever the outcome. The promise of Wi-Fi didn't materialise, the TV channels were limited, and my phone coverage was non-existent. I could not check the news and had to make do with my imagination and the scene playing out before me. As the wind howled, the ship creaked and groaned. The spray from below reached my seventh-floor window, reminding me what the sea was capable of. The sturdy boat built for these seas was certainly coping with the challenge, but it was by no means comfortable. I was trapped between hell and high water. Trying in vain to stay in bed, the heavy fruit bowl hurtled across the room, connecting with my head while my suitcase and its contents scattered. The furniture not bolted down rearranged itself. The sinister grey waterfall outside my window saw me praying for my life. The noise of the gale, together with the grinding of the ship's propeller, was paralysing. I prayed to every known deity that if they brought me safely to the shores of Denmark, I would be a better person.

A knock at the door and I dreaded the worst. It was the ship's purser advising me to vacate this cabin. They were closing the top decks for safety, and I had to move to a lower deck. I felt as though I hadn't blinked my eyes for the remainder of the voyage but held on tightly and waited for the end. Finally, it was over. Land lay ahead, and the final hour was calm as MV Norröna inched her way towards Hirtshals.

By the time I disembarked, my entire body ached. My mind was wretched from fear, and my eyes gritty from not blinking and lack of sleep. I made my way down the companionway to the car deck. It was obvious that not all passengers had good sea legs, and I didn't envy the cleaner's job. I pulled my scarf tighter around my nose and mouth and proceeded quickly to my car. Driving off, I glanced back to see the cracked bridge windows, the only visible damage to the ship.

Is a crossing on the Norröna in winter a good idea? Well, if, like me, you wished to take your car to Iceland, then it was your only option. I would say this. If I found myself on a ship in winter, it was the flagship of the Smyril Line M/V Norröna that I would trust my life with. She was well built, capable, sturdy, and safe. Would I do the journey again? Never!

PS: SURVIVAL

On Tuesday 24th March 2020, the Icelandic Health Authorities confirmed their first death in Iceland from Covid-19. Following this event, and just like the rest of the world, Iceland imposed bans on gatherings of people, travel, working practices, and public events. Museums, entertainment venues, schools, universities, and worse still, the swimming pools were all closed for varying periods of time.

Covid-19 was not Iceland's first pandemic. A previous pandemic back in 1918 catapulted the use of geothermal activity into the daily lives of the Icelandic population. That was the Spanish Flu. During an unusually cold winter and combined with sky-high coal prices due to the war, Iceland's fragile economy had to find alternative ways to produce affordable heating. Harnessing the abundance of natural energy was their solution. Today, around 65 per cent of Iceland's energy supply is geothermal. Icelanders are survivors.

When I think how my own journey became dominated by the swimming pools and hot pots and the purpose they serve in the community, it is almost unimaginable to think of them being closed. Images of Icelanders forming long queues to resume their love of taking a dip were reported in the news in the same way the UK queued for a pub or hairdresser.

Finally, after two years of restrictions, I travelled back to Iceland to pick up where I had left off. One of the four families I had hoped to interview had lost elderly relatives, and with their loss, the stories they had hoped to share. Others, thankfully, were healthy, well and eager to answer the many questions I had surrounding the occupation of Iceland during the war years and "The Situation".

PPS: ESSENTIAL ITEMS

One of my most difficult challenges was staying warm after a hike or activity that caused me to sweat. As my sweat froze, I was immediately cold. I finally admit that wool is best. It beats poly-anything.

- Three sets of Icebreaker Merino base layers of varying thickness and not too tight. It's the air that traps between the layers that insulate you.
- Wind and Waterproof. I repeat <u>wind</u> and waterproof. I packed two lightweight jackets so I could wear them together if needed.
- Spare gloves, hat, and neck warmer. I had three pairs of liner gloves, one thin and two medium thicknesses. I also had my Hestra Expedition gloves with their quilted liner (and missed them after losing them on the cliffs whilst saving my own life).
- I invested in a pair of USB hand warmers and a few packs of single-use crystal feet and hand warmers. They proved invaluable.
- HanWag walking boots and pull-on diamond ice crampons. On this trip, I didn't have my own insulated overall though after loaning one several times, I have since invested in my own and take it everywhere I can. I bought Regatta Lifeguard from Norway.
- Flip flops for the hot pots and pools (*useless in winter*).
- One pair of Devold "Nansen" socks and several pairs of wool power liner socks.

- Three quick dry towels, one for body, one for hair, and one spare.
- Two thermos flasks (one with a wide opening for soups and hot dogs).
- Swimsuit – take any old one as it will be ruined from algae and salt.
- Bra – not underwired, as they really are colder. I bought Icebreaker, which were warm and dried quickly when washed.
- First Aid Kit – including an elastic bandage, extra plasters, throat pastilles and a general-purpose painkiller.
- Camera and lenses are a personal choice. I have a Nikon D850 (FX) and a Nikon 7100 (DX). I used two Nikon lenses with the FX camera (AF-S 24-70 /AF-S 200-500) and three lenses with my older D7100 (35mm F1.8G AF-S, an 18-300mm F3.5-6.3G AF-S and a 10-24mm f3.5-4.5 G AF-S). During the trip, I invested in an additional 14-24mm F2.8 lens, mainly for aurora photography.
- Extras. MB-D18 battery pack plus lock, SB-5000 speedlight, EN-EL18c L ion battery, Sony UXD memory card.
- Sturdy tripod.
- Camera bag – I love my ONA shoulder bag, but many people prefer a backpack. I find a shoulder bag easier to access with heavy clothes on.
- Sunglasses – especially in the winter when the sun is low with the added glare of ice and snow. SPF sun and wind protection plus lip balm.
- A roll of dog poop bags – when you are caught short, pick up after yourself. There are almost no toilets in the countryside.
- Hilleberg all season tent (Staika). Expensive but a good investment.
- Sleeping bag – Fjällraven down filled Polar minus 20.

- Therm A Rest mattress and memory foam pillow.
- Mini "Trianga" kitchen set with spare gas flask, kettle, and Primus gas burner.
- Headlamp with backup power pack.
- Knives – I had several and kept one in each pair of trousers, so was never without one.
- A few miniatures of whisky from home. Ideal for saying thank you.
- Soft hair scrunchies and Philip Kingsley Elasticizer. Good enough for Audrey Hepburn, good enough for me.

Apart from the Lundhags boots, I used everything.

PPPS: RULES OF THE ROAD

- Vegagerðin.is app is your lifeline.
- If you are advised a road is impassable, believe it.
- Remember the locals have grown up with the Icelandic driving conditions, yet even they have accidents.
- The 112 Emergency app is available for everyone to download, and it's a great idea. It updates your progress and sends coordinates to ICESAR. Do remember to sign out of it though when you have arrived at your destination.
- Service stations can be few and far between outside of the larger towns. Fill up with fuel when you can and make sure you have water, snacks, and if it's winter, ideally a thermos in the car with you.
- Watch out for livestock, especially sheep, lambs and even horses and the odd reindeer.
- Think when you stop for photos how your car impacts other drivers. Be considerate and safe. Remember, there are no trains in Iceland. Heavy freight goes by road and air.
- Use the loo at the service station or tourist sites where possible. If that's not possible, remember to pick up after yourself. Iceland sees way too much shit left behind.

© Emma Strandberg